St Antony's Series
General Editors: **Jan Zielonka**, Fellow of St Antony's College, Oxford and **Othon Anastasakis**, Research Fellow of St Anthony's College, Oxford, and Director of South East European Studies at Oxford.

Recent titles include

Nayef R. F. Al-Rodhan
META-GEOPOLITICS OF OUTER SPACE
An Analysis of Space Power, Security and Governance

Carla L. Thorson
POLITICS, JUDICIAL REVIEW AND THE RUSSIAN CONSTITUTIONAL COURT

Daisuke Ikemoto
EUROPEAN MONETARY INTEGRATION 1970–79
British and French Experiences

Nayef R. F. Al-Rodhan
THE POLITICS OF EMERGING STRATEGIC TECHNOLOGIES
Implications for Geopolitics, Human Enhancement and Human Destiny

Dimitar Bechev
CONSTRUCTING SOUTH EAST EUROPE
The Politics of Balkan Regional Cooperation

Julie M. Newton and William J. Tompson (editors)
INSTITUTIONS, IDEAS AND LEADERSHIP IN RUSSIAN POLITICS

Celia Kerslake, Kerem Ŏktem and Philip Robins (editors)
TURKEY'S ENGAGEMENT WITH MODERNITY
Conflict and Change in the Twentieth Century

Paradorn Rangsimaporn
RUSSIA AS AN ASPIRING GREAT POWER IN EAST ASIA
Perceptions and Policies from Yeltsin to Putin

Motti Golani
THE END OF THE BRITISH MANDATE FOR PALESTINE, 1948
The Diary of Sir Henry Gurney

Demetra Tzanaki
WOMEN AND NATIONALISM IN THE MAKING OF MODERN GREECE
The Founding of the Kingdom to the Greco-Turkish War

Simone Bunse
SMALL STATES AND EU GOVERNANCE
Leadership through the Council Presidency

Judith Marquand
DEVELOPMENT AID IN RUSSIA
Lessons from Siberia

Li-Chen Sim
THE RISE AND FALL OF PRIVATIZATION IN THE RUSSIAN OIL INDUSTRY

Stefania Bernini
FAMILY LIFE AND INDIVIDUAL WELFARE IN POSTWAR EUROPE
Britain and Italy Compared

Tomila V. Lankina, Anneke Hudalla and Helmut Wollman
LOCAL GOVERNANCE IN CENTRAL AND EASTERN EUROPE
Comparing Performance in the Czech Republic, Hungary, Poland and Russia

Cathy Gormley- Heenan
POLITICAL LEADERSHIP AND THE NORTHERN IRELAND PEACE PROCESS
Role, Capacity and Effect

Lori Plotkin Boghardt
KUWAIT AMID WAR, PEACE AND REVOLUTION

Paul Chaisty
LEGISLATIVE POLITICS AND ECONOMIC POWER IN RUSSIA

Valpy FitzGerald, Frances Stewart and Rajesh Venugopal (editors)
GLOBALIZATION, VIOLENT CONFLICT AND SELF-DETERMINATION

Miwao Matsumoto
TECHNOLOGY GATEKEEPERS FOR WAR AND PEACE
The British Ship Revolution and Japanese Industrialization

Håkan Thörn
ANTI-APARTHEID AND THE EMERGENCE OF A GLOBAL CIVIL SOCIETY

Lotte Hughes
MOVING THE MAASAI
A Colonial Misadventure

Fiona Macaulay
GENDER POLITICS IN BRAZIL AND CHILE
The Role of Parties in National and Local Policymaking

Stephen Whitefield (editor)
POLITICAL CULTURE AND POST-COMMUNISM

José Esteban Castro
WATER, POWER AND CITIZENSHIP
Social Struggle in the Basin of Mexico

Valpy FitzGerald and Rosemary Thorp (editors)
ECONOMIC DOCTRINES IN LATIN AMERICA
Origins, Embedding and Evolution

Victoria D. Alexander and Marilyn Rueschemeyer
ART AND THE STATE
The Visual Arts in Comparative Perspective

Ailish Johnson
EUROPEAN WELFARE STATES AND SUPRANATIONAL GOVERNANCE OF SOCIAL POLICY

Archie Brown (editor)
THE DEMISE OF MARXISM-LENINISM IN RUSSIA

Thomas Boghardt
SPIES OF THE KAISER
German Covert Operations in Great Britain during the First World War Era

Ulf Schmidt
JUSTICE AT NUREMBERG
Leo Alexander and the Nazi Doctors' Trial

Steve Tsang (editor)
PEACE AND SECURITY ACROSS THE TAIWAN STRAIT

James Milner
REFUGEES, THE STATE AND THE POLITICS OF ASYLUM IN AFRICA

Stephen Fortescue (editor)
RUSSIAN POLITICS FROM LENIN TO PUTIN

St Antony's Series
Series Standing Order ISBN 978–0–333–71109–5 (hardback)
978–0–333–80341–7 (paperback)
(*outside North America only*)

You can receive future titles in this series as they are published by placing a standing order. Please contact your bookseller or, in case of difficulty, write to us at the address below with your name and address, the title of the series and the ISBNs quoted above.

Customer Services Department, Macmillan Distribution Ltd, Houndmills, Basingstoke, Hampshire

---

Meta-Geopolitics of Outer Space

An Analysis of Space Power, Security and Governance

Nayef R. F. Al-Rodhan

Senior Member, St Antony's College, Oxford University, UK and Senior Scholar in Geostrategy, and Director of the Geopolitics of Globalisation and Transnational Security Programme, Geneva Centre for Security Policy, Geneva, Switzerland

palgrave macmillan

In Association with St Antony's College, Oxford

First published 2012 by
PALGRAVE MACMILLAN

Palgrave Macmillan in the UK is an imprint of Macmillan Publishers Limited, registered in England, company number 785998, of Houndmills, Basingstoke, Hampshire RG21 6XS.

Palgrave Macmillan in the US is a division of St Martin's Press LLC, 175 Fifth Avenue, New York, NY 10010.

Palgrave Macmillan is the global academic imprint of the above companies and has companies and representatives throughout the world.

ISBN 978–0–230–31445–0

This book is printed on paper suitable for recycling and made from fully managed and sustained forest sources. Logging, pulping and manufacturing processes are expected to conform to the environmental regulations of the country of origin.

A catalogue record for this book is available from the British Library.

Library of Congress Cataloging-in-Publication Data
Al-Rodhan, Nayef R. F.
Meta-geopolitics of outer space / Nayef R.F. Al-Rodhan.
pages cm
Includes bibliographical references and index.
ISBN 978–0–230–31445–0
1. Geopolitics. 2. Outer space—Political aspects. 3. Astronautics—Political aspects. 4. Astronautics and state. 5. Technology and international relations. 6. Space security. 7. Space control (Military science) 8. Outer space—Law and legislation. 9. Outer space—International cooperation. I. Title.
JC319.A4938 2012
320.1'20919—dc23 2012009546

10 9 8 7 6 5 4 3 2 1
21 20 19 18 17 16 15 14 13 12

Transferred to Digital Printing in 2012

Contents

List of Tables

Tables

Acknowledgements

The author would like to thank the following people for their help: Amber Stone-Galilee, Liz Holwell, Andrew Mash, Bethany Reichenmiller, Beatrice Fihn, Julia Knittel, Frederique Guerin and Anna Koroleva.

He also would like to thank his colleagues at St Antony's College, Oxford University, and the Geneva Center for Security Policy, Geneva, Switzerland, for their help and support.

The author and publishers wish to thank the following for permission to reproduce copyright material: LIT Publishers, for the Figure 4.1, from Nayef R. F. Al-Rodhan (2007) *The Five Dimensions of Global Security: Proposal for a Multi-sum Security Principle* (Berlin: LIT).

Every effort has been made to trace rights holders, but if any have been inadvertently overlooked, the publishers would be pleased to make the necessary arrangements at the first opportunity.

The views expressed in this book are entirely those of the author and do not necessarily reflect those of St Antony's College, Oxford University, or the Geneva Center for Security Policy.

1
Space Power: An Introduction

This volume is an attempt to shed new light on the current debate about the highly complex and fast-changing geopolitics of space. Unlike many existing works examining current space technologies and national space policies, this book goes beyond traditional international relations approaches to space by applying the comprehensive framework of *meta*-geopolitics to space. One of the most distinctive traits of the *meta*-geopolitics framework is its all-encompassing view of power, security and geopolitical actors. The *meta*-geopolitics approach does not undermine traditional, geography-related factors in the current geopolitical analysis of space but it adds new dimensions of power and thus broadens its definition. The *meta*-geopolitics framework highlights seven capacities through which the power of a state or a non-state actor can be assessed: (a) social and health parameters; (b) domestic politics; (c) economics; (d) the environment; (e) science and human potential; (f) military and security factors; and (g) international diplomacy. As is set out in my previously published work, *Neo-Statecraft and Meta-Geopolitics: Reconciliation of Power, Interests, and Justice in the 21st Century*,

> *Meta*-geopolitics differs from traditional concepts of geo-politics, as it proposes a multidimensional view of power. It includes all of the soft- and hard-power tools that states can employ to project power. Moreover, it assesses the relative strengths and weaknesses of each capacity. The *meta*-geopolitical approach also allows us to make predictions about the ability of a state to continue to project its power in the future. It takes into account variables like projected population growth, demographic make-up, public health and human and scientific potential. These factors help us gauge whether a country will retain, increase or lose its current geopolitical status.[1]

The advantage of the *meta*-geopolitics framework is that the above capacities not only determine the geopolitical status on Earth but they can also be directly applicable to space. In comparison with traditional geopolitical approaches, deeply grounded in geography, demography and economics, the *meta*-geopolitics framework includes space and sheds a new light on the interplay between space powers and geopolitics on Earth. Furthermore, while this analytical grid is relevant for states it also applies to private-sector entities, which are playing an increasingly important role in contemporary space matters.

This book is an attempt to respond to the need for a paradigmatic shift in today's globalized, interdependent and interconnected world, whose activities have spread far beyond the Earth's surface to enter the realm of space. On the one hand, it aims to make an analytical contribution to the field of the geopolitics of space by offering an innovative, multidimensional *spatial* framework to account for the new features of space power. It also strives to clarify the link between activities in space and their influence on international politics on Earth. *The Meta-Geopolitics of Outer Space* could therefore be defined as a 'grounded', 'elevated' and 'connecting' framework providing new insights into how global actors interact with each other in space, the impact of space technology on their respective positions on Earth and ultimately what type of governance they may seek to develop for space. This model enables a comprehensive assessment of the strategic orientations of major and emerging space powers. Such a complete analysis is crucial for identifying new threats, making informed decisions and devising multilateral solutions to common space security challenges.

The unique opportunities of space

Today, there are more states building up a space presence, more potential avenues for space exploration and more international industries devoted to space and space-related technologies than ever before. It is estimated that '[t]here are over 1,100 active systems in orbit and an additional 21,000 pieces of debris' orbiting the Earth.[2] Because of its zero-gravity environment and its physical relation to the Earth, outer space offers a forum for many activities, endeavours and scientific research projects that would be impossible to conduct from the ground or from within the Earth's atmosphere. The unique functions and opportunities range from the environmental, economic and developmental to matters of diplomacy and national prestige, social and health applications and, of course, military uses. Environmental monitoring of climate change, deforestation or changes in atmospheric behaviour, advanced global communication systems, weather forecasting and weather monitoring, improved position determination and navigation, advanced reconnaissance and early warning systems, astronomical studies, as well as various health- and illness-related experiments are just a few of the countless possible applications of outer space.

Moreover, the numerous opportunities that outer space presents are constantly evolving in the light of new and more advanced space technologies. Whether it is improved satellites' protection or more efficient designs for astronauts' clothing, constant innovations in space technology create even more ways for businesses and governments to utilize space for profit and/or security. Importantly, many of the technological developments driven by space exploration and research also have a huge impact on our day-to-day lives on Earth. Outer space offers opportunities for learning and observing, and for commercial profit but, perhaps most importantly, outer space and human engagement with it are changing the way in which individuals, countries and the international community think about their physical, intellectual and human security priorities.

While the changing situation in space opens up fascinating new avenues and opportunities, it also presents insecurities, threats and challenges. Exponential developments in technology, the decreasing cost of space access and the unprecedented use of space for commercial purposes mean that space is becoming more and more crowded. Congestion brings threats such as satellite collision and disputes over the allocation of orbital assignments, not to mention the hundreds of thousands of pieces of dangerous space debris such as meteorites or wreckage from retired satellites that are moving at high speed around the Earth.[3] Moreover, the heavy reliance on space applications makes individuals, countries and businesses increasingly vulnerable to potential disruptions and threats. Or, as a representative for Philip J. Baines puts it, '[a] day without space would be a disaster. The next hundred years without space would be a catastrophe.'[4]

The potential weaponization of space presents a serious geostrategic challenge to the international community. The growing number of space-faring players has resulted in an increasingly complex and competitive geopolitical situation in which the needs of countries to protect their national interests, pride and security may ultimately lead them to introduce weapons into space. If countries such as the United States, Russia or China decide to use their space presence as a sign of their control over this realm, the result could be a global space arms race.

In the context of these new opportunities and threats, questions arise over security, governance and power relations in space. How is it possible to make the best use of promising new opportunities, and at the same time contain existing and potential future threats? What dimensions should policymakers consider when formulating space policy? What are the main considerations that need to be taken into account in future treaties and international agreements governing space? Is collaboration in space possible? How can we secure outer space for all? Is space weaponization inevitable? In order to answer these questions, policymakers and other decision makers are in dire need of explanatory frameworks for the politics and geopolitics of outer space.

This introductory chapter discusses classical geopolitics and its pertinence to the complexities of space. It examines the way in which the three major schools of international relations, realism, liberalism and critical geopolitics have been applied to space and to space security issues. It then challenges these approaches and makes a strong case for a new approach: *meta*-geopolitics of outer space including its related paradigms of '*symbiotic realism theory*' and the '*multi-sum security principle*'.

Existing geopolitical approaches to space

As the list of possibilities for utilizing outer space increases, along with greater international access to and interest in space, the geopolitical context in which space policies are carried out is profoundly changing. This shift first began in the 1990s after the break-up of the Soviet Union, but it continues in full force today. In the 1950s, only a handful of countries had a space presence or even an active space research programme. Today, more than 60 countries have at least some presence in space, even if it is only a weather satellite or sending researchers to the International Space Station (ISS).[5] Private and commercial actors have also discovered the great potential of space flight and technology, and are gaining more and more access to space, thereby influencing and altering geopolitical power configurations.

The complexity of this constantly evolving situation, the multiplicity of actors and the speed of technological change pose serious challenges to any policymaker involved in space matters. In order to assess, analyse and deal with current and future power relations in space, statesmen need to get a clear picture of the relative position of their country to other international actors. They need to be aware of their country's capacities, be they military, economic, environmental or any other, in order to be able to make informed decisions and choose appropriate policies and tools to protect their citizens and defend their interests.[6] The literature on geopolitics has traditionally sought to provide answers to such questions and tools to explain and assess such configurations. For the purposes of this book, the concept of classic geopolitics is defined as

> the study of how geography, economics and demographics impact politics, particularly political relationships among states. It looks at the power dynamics among states seeking to control territory and to acquire reliable access to strategically important locations and resources.[7]

Geopolitics is a highly controversial subject and has been debated extensively since its inception by the Swedish political scientist Rudolf Kjellén in 1899. It has never been a formal theory, but instead only an 'attempt to reveal textually and cartographically the complex relationships between geography and politics at a variety of spatial scales from the local to the global'.[8]

The traditional view of geopolitics has been heavily criticized. It was most popular around the beginning of the twentieth century. Traditional geopolitics examines the potential of geographical factors to influence and determine the power, productivity and history of a specific geographical area.[9] The context in which it first emerged was characterized by great economic and political changes. At the end of the nineteenth century, a new model of capitalism and a new world order led by the United States emerged, as electricity, gas, oil, Fordist types of production and expanding mass markets became prominent.[10] New communications technologies such as the telegraph and new modes of transport such as the railways contributed to make the world a more interconnected and interactive place. As a result, one of the main questions addressed by early geopoliticians was the effects of new technologies on international politics.[11] Half a century later, geopolitical theory was taken to the extreme by Nazi geopoliticians such as Karl Haushofer.[12] Such approaches discredited the concept of geopolitics for their determinism and abuse for imperialist reasons.

In the 1950s, however, geographic factors were taken up again by early classical realists, who re-introduced them into the domain of international politics. Theorists such as Hans Morgenthau and Nicholas Spykman restored the reputation of geopolitics somewhat by placing more emphasis on human action and by limiting the influence of geography to the domain of foreign policy – rather than applying it to almost all human activity as previous geopoliticians had done. Another change was that, instead of examining geography alone, they introduced additional factors such as natural resources, industrial developments and social institutions.[13] As a consequence, its reputation was re-established and the perceived importance of traditional geopolitical factors such as geography continued to shrink.

As the world became more globalized, and with the development of advanced technologies for air and space power, space has emerged as another determinant or criterion of political relationships. When the Soviet Union launched Sputnik, the first satellite in space, Senator Lyndon B. Johnson declared that 'whoever controls space controls the world'.[14] Fifty years later, the increased reliance on space technologies, the large number of new actors in space and the commercial value of space exploration demonstrate the importance of space to national politics in the twenty-first century. Furthermore, given the direct correlation between activities in space and those on Earth, international actors can no longer ignore the role of space in national security issues, the new opportunities it opens up and the challenges it presents, and the centrality of space in defining contemporary geopolitical configurations.

Although the threats and opportunities of space use are capable of affecting our daily life on Earth, the debate about space has mostly been conducted among highly specialized experts, and has been widely ignored by the public.[15] These debates have not resulted in the emergence of a single comprehensive geopolitical theory of space, but they have generated a number of controversial schools of thought. Many among them are applications heavily influenced by existing international relations paradigms ranging from classical theories, such as realism, neo-realism, liberalism and neoliberalism, to more recent approaches including social constructivism and gender theory.[16] To what extent have the different paradigms of international relations influenced our understanding and interpretation of space and contributed to the emergence of practical geopolitical theories of space? What are the strengths and weaknesses of these approaches and how useful are they for policymakers or space-faring actors today?

The influence of realism

Early explanations of the geopolitics of space rely heavily on classical realist assumptions. This is not surprising since realism was the predominant approach in international relations in the 1950s, when the Soviet Union launched the world's first satellite into space and the Cold War was at its height.[17]

Classical realism, as popularized by Hans Morgenthau, considers states to be unitary and rational actors. In the absence of a world-recognized authority, territorial states compete for power in an anarchic global system. Competition is considered the only way to survive and to maximize a state's unilateral interests.[18] US and Soviet efforts to explore and use space were therefore considered a by-product of rival super-powers striving for power and military capacities. As Michael Sheehan points out, a space programme was regarded as a valuable contribution to overall power.[19] In the realist view, the unilateral striving for power to improve national security may lead to a security dilemma, a spiral of insecurity and pursuit of security caused by the perception of the other power's efforts to accumulate power.[20] From this perspective, the notion of 'independence' appears to be crucial and space is perceived as 'a new frontier that will be exploited as part of an inevitable and enduring struggle for power'.[21] During the Cold War, the realist theory proved to be relevant to the analysis of the US-Soviet competition in space and the national security issue.[22] In the late 1970s, the neo-realist school of thought emerged and proved to be useful in a changing international space environment. The balance of power politics is a recurring theme for neo-realists. Kenneth Waltz, who is widely regarded as the founding father of neo-realism, maintains that states in an anarchic international system tend to counterbalance a rising power. Rather than seeking to join the new power, they will form alliances in order to face the threat they see as emerging.[23] With the emergence of new space powers, including the European Union (EU), China and India, neo-realism saw its premise confirmed as the space order started to shift from a proposed bipolarity to a multipolarity of powers.

Everett C. Dolman's *Astropolitik* offers an attempt to apply key assumptions of realism to outer-space politics. In his book, the author aims at 'corralling the elements of space and politics recognized as realist into their proper places in grand strategy'.[24] Dolman proposes an *Astropolitik* model which is defined as a 'determinist political theory that manipulates the relationship between state power and outer-space control for the purpose of extending the dominance of a single state over the whole of the Earth'.[25] Deeply attached to realist tenets, Dolman argues that a competitive quest for material benefits should maximize space exploration, which in turn will benefit all people. Furthermore, Dolman holds that every type of cooperation is originally motivated by competitive considerations and fear. Space is seen as a yet unpossessed resource which is to be conquered and exploited by mankind.[26]

Fully adhering to the *space nationalist* school of thought, Dolman also applies realist postulates to the field of space security.[27] As a result of an anarchic international system and competition, 'the reality of confrontation in space politics pervades the reality of the ideal of true cooperation and political unity in space which has never been genuine, and in the near term seems unlikely'.[28] According to Dolman, all attempts to limit and regulate weapon use and proliferation in space are merely 'a slick diplomatic maneuver'.[29] Hence, space-related treaties and international agreements are unlikely to work if engaged states can withdraw from them when they feel threatened.[30] Space nationalists hold that what was considered as cooperation during the Cold War was in reality 'the result of a lack of state interest in pursuing the types of space defenses or other military activities limited by space-related treaties and agreements' they had signed, such as the 1967 Outer Space Treaty (OST).[31]

A more extreme branch of space nationalism, inspired by Dolman's work, goes even further in its application of realism to space and calls for an extreme version of space nationalism – 'US space hegemony'.[32] Dolman argues that as long as the world is not democratic, unilateral hegemony in space will remain the sole means to ensure peace and prosperity for all. His approach presumes that the 'state that dominates space is specifically chosen by the rigors of competition as a politically and morally *superior* nation, culture, and economy'.[33] Therefore, as the world's dominant liberal democracy, the United States must seize control over space and guide all other nations in the use of space.

Based on these considerations, Dolman proposes a blueprint for a US space policy or an *Astropolitik* in three steps:

(1) The United States should withdraw from the existing space regime, namely, the OST. Instead of the 'global commons' approach at the heart of the OST, Dolman advocates a 'free-market sovereignty in space'.[34] He argues that a new space regime should be introduced, which is structured like a free-market economy and allows for the privatization of space and competition for profit.[35]
(2) The United States should deploy a space-based Ballistic Missile Defence (BMD) system which would enable the military control of low-Earth orbit.[36]
(3) A specialized US space coordination agency should be established.[37]

By implementing this strategy, Dolman believes that 'a dominant liberal democracy like the United States can usher in a new era of peace and prosperity'.[38] However, while Dolman is convinced about the benefits *hegemony* would bring to international peace and security, he seems to be more uncertain about the feasibility of such a hegemonic regime in the existing geopolitical space environment. In his *Final Thoughts* about *Astropolitik* Dolman clarifies that this strategy is 'an examination of optimal strategies and likely outcomes given an assumption of near-term continued nationalist military and economic competition [...], it is not a prediction or a portend of probable outcomes'.[39] In fact, given the increasing number of actors entering space and the growing interconnectivity between them, the emergence of a hegemonic space power, as advocated by Dolman, seems to be unlikely and *Astropolitik* fails to provide a realistic picture of existing and future power relations in space.

There are many other academic and practical criticisms of this approach and a substantial number of them stem from advocates of the critical geopolitical view. They are discussed in more detail below. Suffice to say here that one of their most common critiques is that Dolman offers an explanation from an imperialist view, but fails to theorize the normative implications of such a unilateral space-based dominance.[40] Other criticisms concern the more general nature of the realist approach. A realist geopolitical approach to space risks neglecting the importance of the ever-increasing number of non-state actors in space. As private companies and even individuals gain easier access to space, their influence on geopolitics needs to be taken into account. Moreover, the many non-military threats to security (discussed in more detail in Chapter 3) are not given due importance. Elsewhere, I have argued that '[c]hange and transformation occurs not simply because of the distribution of material power (military or economic) inside the system, but also as a result of non-material forms of power. Notably, cognitive-normative schemes also have an impact on the dynamics of the international system and these are missing in the realist paradigm.'[41]

Moreover, realists consider that cooperation is unlikely to happen but, if it does, it is believed to be the result of rational calculations and a security dilemma. Alternative views suggest that the history of space flight has witnessed many examples of what appeared to be cooperative acts, and states may use cooperation as a means to soften tensions in their relationships. Examples cited include the project of space cooperation between the United States and the Soviet Union during the Cold War and between China and Russia.[42]

The liberal alternative

The second predominant model for conceiving of the geopolitics of space is *the republican/liberal school*. Liberalism distinguishes between a state of war and a state of peace, rather than assuming that the world is in a permanent state of war caused by competition between states.[43] Its central and most important principle is the freedom of the individual and its right to moral treatment. Institutions safeguarding the rule of law need to be established to maintain these basic rights. Neoliberalism, as described by Robert Keohane, also considers the creation of such institutions to be crucial. Rational and self-interested states are the principal actors, but international regimes need to be set up in order to facilitate cooperation and avoid market failure.[44] International law, human rights and free trade need to be guaranteed by such international institutions.

Liberalism and neoliberalism provide several useful elements of explanation for the dynamics of outer space. In contrast to realist approaches, both liberalism and neoliberalism allow for non-state actors and the domestic considerations of space-faring states to be taken into account. Furthermore, because neoliberals believe that the military use of space needs to be constrained by the promotion of cooperation and the establishment of international regulatory structures,[45] neoliberalism lends itself to the analysis of the international law of space, the OST and multilateral institutions such as the European Space Agency or the United Nations.[46] The liberal view of space has been particularly useful since the end of the Cold War, when multiple new space powers such as the EU, China and India have emerged, the uses of space have diversified into military and civilian applications and more cooperation could be observed.

One of the most important writers of the school of *liberal astropolitics*, or *liberal-republican astropolitics*, is Daniel Deudney. Like Dolman, Deudney believes that military control of space by one state could lead to the domination of the entire world.[47] Yet '[r]ather than developing the implications of this as a strategic opportunity for any one state (e.g. the US) [...] Deudney sees it as a collective problem to be kept in check through collaboration'.[48] Joint efforts, rather than unilateral dominance of space, may contain security dilemmas. While facing common security challenges, ties between states will be strengthened and sovereignty will nevertheless be maintained.[49] In other words, Deudney calls for what has been termed the '*astropolitics of collaboration*'.[50] He shows that there have been many examples in the history of space flight that provide proof of productive collaboration resulting in stronger and more stable relationships between the states involved.[51] One such example is the Sputnik mission itself, which had initially come into being as an internationally supported research programme – the International

Geophysical Year. A more contemporary example is the International Space Station, which is jointly used and supported by Russia, the EU and the United States.[52] As a consequence, Deudney regards institutionalized cooperation as a way of promoting peace. In other words, he 'believes that geopolitical attention to material conditions, spatiality, change, and political processes could form the basis of a theoretically sophisticated contextual-materialist security theory of world politics'.[53]

Deudney's concept of *institutionalized cooperation* is shared by the authors of the *global institutionalism* school, which represents, with space nationalism, one of the largest schools of thought in the field of space security. *Global institutionalism* emerged in the late 1950s. Initially, it drew on idealist tenets from Grotius and Kant before turning, in the 1980s, towards more realistic neoliberal postulates.[54] Global institutionalists highlight the importance of space treaties, international institutions and any other form of bilateral or multilateral cooperation in preserving space benefits and thereby contributing to peace on Earth.[55] Firmly opposed to the realist model and space nationalist school, global institutionalists see in space a unique opportunity to escape from international anarchy, and in cooperation a 'form of enlightened self-interest'[56] and 'improved behaviour'[57] of states that would in the end benefit mankind. Whereas space nationalists see in the signing of the OST 'merely a reaffirmation of Cold War realism and national rivalry',[58] from a global institutionalism perspective it is considered a major common achievement in the field of weapon regulation and the non-proliferation of arms in space.[59]

As an alternative to the institutionalized cooperation presented by Deudney, the *optimistic school of technological determinism* proposed a 'science-based "convergence"' theory.[60] Instead of taking politics as its starting point, this approach focuses on technologies, the development of which will drive further socio-political changes on Earth. In the domain of space, the technological determinists believe that 'cooperation in space would arise out of the objective forces of advanced scientific research and development'.[61] The high costs and complexity of space technologies provide a strong incentive for states to cooperate in that field.[62] Furthermore, the financial and technological benefits resulting from this collaboration would foster convergence even between ideological rivals, thereby reducing existing political and ideological divergences and contributing to international stability.[63]

Another technological determinist variant, influenced by the European school of political economy and the theory of '*collective goods*', predicts that given the increasing number of actors in space, coupled with the commercial as well as military benefits resulting from their activities, cooperation is no longer in the interests of space-faring nations.[64] As the Swedish analyst Per Magnus Wijkman points out, however, this increased presence has created new interconnections among different players in space, such as the growing danger of collision. In the light of these developments, self-interested rational actors might engage in collaborative dynamics at least 'to keep interference at a mutually acceptable level'.[65]

In sum, all the variations of optimistic technological determinism seem to agree on the need for and general benefits of collaborative behaviour among space-faring actors. Although they differ on the factors that ought to lead to this collaboration, most agree that 'management arrangements may be possible but will rely on favourable structural conditions, communications and political bargaining'.[66]

It is also worth mentioning another school of space security, *social interactionism*, which proposes an alternative scenario of cooperation in space. Although it agrees with most of the assumptions of global institutionalism it sees soft tools and voluntary efforts, rather than binding legal regimes, as driving activities in space. One of the school's early thinkers, Paul Stares, proposes a *rules of the road* approach.[67] Contemporary representatives of this school, such as Michael Krepon and Michael Katz-Hyman, explain that given the complexity of political reality there are situations where soft law rules 'may be easier (and quicker) to obtain among the multiple players in space, some of which are no longer nation-states',[68] rather than binding international treaties. Both authors are favourable to the development of codes of conduct and call on the United States to take the lead in the design of the new space regime.[69]

Today, liberal and neoliberal ideas are widespread throughout the international community. The need for genuine cooperation and robust international tools to monitor the use of space, and thus preserve international peace, has been universally acknowledged in recent decades. The neoliberal school of international relations has undeniably added value to the geopolitics of space in terms of its analysis of international institutions and space-related regulation mechanisms.[70]

In addition, the liberal approach is particularly relevant in analysing the impact of domestic politics on national space programmes. As Sheehan explains, liberalism considers national space programmes not as the unilateral decisions of a national government, but rather as a result of a political bargaining process among different domestic actors.[71]

Furthermore, this approach considers other factors such as '[h]istory, culture, value systems and "domestic" political structures'[72] and the impact of globalization when analysing national space programmes. As Sheehan stresses, 'the growing cooperation between different national space programmes emphasized the inadequacy of traditionally autonomous states in addressing certain contemporary challenges. The high cost of space ventures, along with the associated technological interdependence makes cooperation both necessary and inevitable'.[73]

Although liberal and neoliberal schools of thought provide interesting insights into space policies, particularly in the post-Cold War context, they nonetheless fail to provide a comprehensive picture of the contemporary geopolitics of space. Like the other existing approaches, the application of liberal postulates to space focuses on only a few dimensions of power. For example, the liberal school was heavily criticized for justifying space programmes only in terms of non-military criteria such as economic and technological progress, thereby neglecting their military component.[74] Similarly, one of the main critiques of Deudney's liberal-republican astropolitical model is that it ignores power relationships.[75] According to Jonathan Havercroft and Raymond Duvall, this cooperative approach leads to a 'symmetrical' system with no single component dominating others. In their view, such a model fails to 'acknowledge the profound asymmetries of aspirations and technological-financial-military capacities among states for control of orbital space'.[76]

The critical geopolitical perspective

The third and most recent school of international relations to heavily influence the field of the geopolitics of space is *critical geopolitics*. Gearóid Ó Tuathail was the first systematic proponent of this new approach, which aims to challenge the traditional geopolitical views.[77] Closely related to post-structuralist and social constructivist postulates, which maintain that 'all knowledge is socially constructed and contingent on human practices informed by collective knowledge and norms',[78] *critical geopolitics* argues that 'power relations are not imposed on already existing subjects: rather, it is within and through power relations that political subjects come into being'.[79]

As applied to space, the social constructivists consider that '[d]ecisions for and against space-related policies, and even decisions about whether to have such policies, are shaped by world views and beliefs about what space does, or might represent'.[80] In a related vein, heavily inspired by *critical geopolitics*, Havercroft and Duvall use its core postulates to challenge the realist and liberal-republican approaches to space. In their *Critical Astropolitics* they propose a revisited theory of space and seek to reformulate the geopolitics of space in a way 'that is more conscious of how power operates in the theory and practice of world politics'.[81]

Havercroft and Duvall question the possible implications of the US space-based hegemony, should it occur, on the structure of the international system. They criticize both Dolman and Deudney for underestimating the impact such hegemony would have on global security with no normative framework and effective counterbalance mechanisms in place.[82] Unbound by traditional boundaries, they argue the weaponization of space would allow the United States to control all states under its hegemony. Moreover, thanks to the precision of space weapons, it would even be possible to target individuals. This option would deprive states of the option to protect their inhabitants and, therefore, affect their sovereignty. US weaponization of space would therefore reorganize the global political order and lead to the creation of a *'space-based empire'* that would erase all boundaries and place the Earth under its total and inalienable control.[83]

One of the biggest contributions of critical geopolitics is that it provides a critical perspective on the mainstream theories of the geopolitics of space by questioning existing power, space and political relations.[84] A large part of this critical literature focuses on the criticism of Dolman's *Astropolitik* and the way it might have influenced the Bush Administration's vision of space. Critical approaches to the geopolitics of space are very useful for casting new light on existing paradigms, identifying their weaknesses and highlighting their limitations. For example, *critical geopolitics* highlights the unique characteristics of space – an endless, yet un-possessed and largely unexplored realm – and the limits of the traditional Earth-bound theories in analysing it. However, this principal strength is at the same time their main shortcoming. Approaches aimed at deconstructing existing paradigms rarely provide new concrete policy guidance and do not produce any guidelines for strategic approaches. Such recommendations are, however, urgently needed to cope with the new challenges posed by the changing nature of space.

The need for a new perspective

Each of the paradigms explored above had its own historical heyday and has had its critics. Yet, maybe their biggest common drawback is that they try to apply Earth-bound assumptions and theories to space and are therefore unable to account for the changes that have occurred since space has become a new area of international competition. The information revolution, communications technologies, mobile telephony, the Internet, the 24/7 news cycle and space imaging technologies available at the click of a mouse have profoundly altered international relations and geopolitics. Today's rapidly changing interconnected and interdependent world, technology, information and communications tools have lessened, but not eliminated, the pivotal role of geography in international relations and political power. New dimensions and actors have emerged that need to be accounted for. Although recent technological developments do not negate the importance of traditional geopolitics and international relations theories, these traditional theories can only remain relevant if new strengths, such as space technology, increased knowledge and human capital, are incorporated into the equation.

In addition to the military applications of space technology, a large number of applications in civil and commercial domains are only marginally taken into account. The threats that countries face today no longer emerge from military uses of space alone. Often, they are decentralized, and potential instabilities emerge from many new areas such as health, the environment, domestic politics or terrorism. The focus of the discussion needs to be broadened and the debate needs to include dimensions that go beyond traditional military and economic factors. Existing geopolitical approaches discussed in this chapter tend to apply traditional assumptions about *terrestrial* power to explain the configurations between international actors in space. Yet, they propose only a limited analysis of the opportunities and threats that space offers today and, more importantly, do not pay sufficient attention to the effects of space assets on Earth.

This book is an attempt to bridge these gaps and provide a new, more comprehensive picture of the geopolitics of space: the *meta*-geopolitics of outer space. It is based on the premise that '[a] spatial analysis of international relations has to consider a number of unevenly distributed resources, or "capacities", to demonstrate the highly complex strategic relationships between states'.[85] Combining traditional and new features of geopolitical analyses, it proposes a multidimensional *spatial* framework, which extends the seven key power capacities listed above into space and analyses how global actors use these space capabilities to improve their respective geopolitical postures on Earth. This multidimensional view of geopolitical power allows consideration of soft, hard, smart and just power tools,[86] offering an all-encompassing perspective on today's complex strategic relationships.

Furthermore, this book seeks to clarify the link between the geopolitics of space and geopolitics on Earth. As the use of space by state and non-state actors expands, space has become an increasingly important component of policy on Earth. In today's instantaneous and interdependent world, space is likely to have important implications for geopolitics and political and societal change. Therefore, understanding the realities of space is crucial to global peace and security, and deserves far more attention in international relations and geopolitics discourses than it currently receives.

Ultimately, this analytical framework aims to provide a tool for the assessment of current geopolitical realities in space, on which concrete policy recommendations can be built. Making policymakers aware of their own and other states' geopolitical capacities is crucial for making strategic decisions in relation to space policy, foreign policy and security policy.

The structure of the book

This book is divided into six chapters. Chapter 2 introduces the framework of *meta*-geopolitics. It sets out its theoretical foundations and discusses its application to space. In it, I also provide my own definition of space power. The chapter briefly explains the way each of the seven dimensions of *meta*-geopolitics could be affected by new space technologies and the possible implications for contemporary geopolitical configurations. Moreover, while *meta*-geopolitics provides the framework of analysis and the structure of the book, this chapter also highlights my previously published approaches to international relations: '*symbiotic realism theory*' and the '*multi-sum security principle*'. Both approaches are closely intertwined with the theory of *meta*-geopolitics and relevant to its space application and governance.

Building on this theoretical basis, Chapter 3 discusses the emergence of space technologies through three major phases: the development of the military, civilian and commercial sectors. It addresses the impact of communications and remote sensing satellites, global navigation systems and space launch rockets on the seven capacities of *meta*-geopolitics. Based on a number of concrete examples, this chapter explains the implications of these new opportunities for state power and strategic considerations.

Chapter 4 discusses space security issues. The chapter investigates existing threats to space security, both intentional and unintentional, and their impact on states' space power capacities and global security.

Chapter 5 applies the theory of *meta*-geopolitics to the national space policies of four space powers (the United States, the EU, China and Russia) and two major non-state actors (EADS and Boeing), highlighting the priorities, strengths and weaknesses of each.

Chapter 6 discusses the main arguments and conclusions of the previous chapters. It reviews existing institutions and legal instruments dealing with space governance and highlights the gaps that need to be addressed. After analysing the space power capacities of state and non-state entities, using the *meta*-geopolitics framework, the implications for geopolitics in the twenty-first century are discussed and potential future trajectories are laid out. Based on this situation assessment, the chapter proposes a comprehensive model for space governance. This relies on the previously published conceptual paradigms of international relations, '*symbiotic realism*' and global security, the '*multi-sum security principle*'.

2
Space Power and *Meta*-Geopolitics

Globalization means that the strategic dynamics of the world we live in are changing rapidly.[1] As suggested previously, '[g]eopolitics is not and never will be a static science. An important part of geopolitical thinking is the study of international dynamics and new developments that might affect the power of states and power relationships between them'.[2] Recent technological progress in transport infrastructures, information and communication tools and, of course, space technologies, has opened new doors but also created additional threats for the international community, thereby altering its existing geopolitical structures.

The classical theories of geopolitics presented in Chapter 1 all define one or a few state capacities – be it military, economic or geographical – as the core element of geopolitics. As is mentioned above, traditional geopolitical factors still matter, reliable access to key resources such as oil, water and industrial infrastructure continue to be key components of an actor's economic development and power. Such assets, today as in the past, have been unevenly distributed around the world, making geopolitics an important part of strategic considerations in the twenty-first century. However, as is explained in Chapter 1, in today's instantaneous and interdependent world, a geopolitical analysis that is limited to classical factors would provide only a partial picture of international power structures. A number of new actors and criteria, such as environmental policies, health, education and diplomatic leverage, influence the status of actors on the international stage. Moreover, these dimensions have spread from the terrestrial to the space realm, making space a critical element of the geopolitical equation.

Globalization and the progress in space technology offer huge opportunities to present and future generations, but have also contributed to the emergence of new challenges. Today, international actors have to face uncertain, diffuse and complex security issues, such as pandemics, transnational threats, energy issues, global warming and space debris. Their sources are numerous and often interrelated, which makes them difficult to identify and address. When dealing with these issues, international actors 'can no longer afford to focus on one particular state or geographically contained area that is the source of security threats'.[3] Instead, they must consider all areas from which potential instabilities and threats could arise and opt for dialogue and coordinated solutions.

Introduction to the *meta*-geopolitics framework

By taking into consideration the complex realities of international relations, the framework of *meta*-geopolitics[4] combines different traditional and new dimensions of geopolitics in order to offer an all-encompassing analysis of the contemporary geopolitical situation. It includes the concepts of classical geopolitics, critical geopolitics (perceptions, representations and interpretations) as well as additional state capacities. It identifies seven areas of power, which are referred to as '*capacities*'. They are social and health issues; domestic politics; economics; the environment; science and human potential; military and security issues and international diplomacy.[5] Often, these components vary not only according to the individual tenets of geopolitics but also within their respective categories. As threats to international peace may emanate from each of the seven capacities, the *meta*-geopolitics multidimensional analysis helps to identify the sources of potential insecurity and thus devise early and tailored responses.

Overall, *meta*-geopolitics deals with power and seeks to provide a nuanced picture of geopolitics in the twenty-first century. If geopolitics seeks to provide a map for statesmen to 'orient themselves in the maze of international relations',[6] *meta*-geopolitics aims to refine this map and make it more precise and more detailed. It aims to help policymakers identify potential threats well in advance, locate vulnerable areas and bring more clarity to an increasingly globalized world. In the introduction to the present work I explain that *meta*-geopolitics is used as a framework throughout this book, with the aim of shedding new light on the debate about the geopolitics of space. This chapter outlines how this framework applies to the domain of space, explains the difference between *meta*-geopolitics and the other approaches to the geopolitics of space and demonstrates the added value of *meta*-geopolitics to global security and space debates.

Towards a broader definition of space power

The multiplication of actors in space, coupled with increased reliance on and use of space technologies, means that space has emerged as an important component of national power and an excellent indicator of the status of an actor in the regional and international arenas. In contrast to the Cold War era, during which states focused on the military uses of space, contemporary actors use space in many other contexts, such as for social and health issues, domestic politics, economics, the environment and international diplomacy.

This section uses *meta*-geopolitics to discuss the notion of *space power*, examines its different components and analyses the role it plays in contemporary geopolitics. The multifaceted approach proposed in this work applies to all types of stakeholders (state and non-state actors). Nonetheless, states remain the most significant players in space, and understanding and defining a state's space power is therefore an important concept.

This section highlights the multidimensional nature of the concept of space power. It argues that its definition cannot be limited to its military capacity, even if this remains important, but must be broadened to encompass all the seven dimensions of state capacities proposed in *meta*-geopolitics. Furthermore, it is argued that space power is a 'two-way' concept because of its ability to influence power dynamics on Earth.

As is noted in Chapter 1, in 1957, after the Soviet Union launched Sputnik, then Senator Lyndon B. Johnson declared 'whoever controls space controls the world',[7] referring to Mackinder's famous declaration 'whoever controls the heartland, will forever seek to dominate the Eurasian landmass and ultimately the world'.[8] The notion of space power was first mentioned in 1964 by Klaus Knorr, but at the time a proper definition was not provided.[9] Ever since, despite many calls for a comprehensive theory of space power, and numerous attempts to conceptualize one, no single, accepted theory of space power has emerged. In the post-Cold War era, a new body of literature on space power began to emerge, particularly in the late 1990s and early 2000s in the United States. In order to explain space power, space theorists and scholars have often evoked parallels with sea power, air power and, to a lesser extent, land power.

Various concepts of land warfare, including the Clausewitzian theory, have been applied to space.[10] For instance, in the military sphere, space is frequently considered to be the ultimate high ground of the global battlefield.[11] This idea is derived from the high ground theory, an often-employed notion in land warfare which states that 'by commanding the hill you will control the surrounding country and can influence the battle in your favor – thereby winning the war'.[12]

Analogies to air and sea power have been used more frequently. Space power is often recognized as being of equal importance for a state's capabilities as the two traditional notions of power. Indeed, space power shares a number of attributes with the notions of sea and air power. Both space and the sea are arenas for commerce, transport, observation and conflict. The establishment of permanent civil and military bases – in orbit or on sea coasts – is important for space flight and sea power.[13] Both are recognized as 'common goods' under international law and as such available for use by all nations. The importance of sea power was first developed by Alfred Thayer Mahan, one of the founding fathers of the geopolitics of the sea, who is commonly considered the first global geopolitician.[14] A convinced proponent of US naval expansion, Mahan theorized that control over coastlines, the possession of significant ports and a strong navy were the key to a state's international power.[15] His work has established itself as a key reference for states' strategic decisions at the beginning of the twentieth century and has shaped decision-making in terms of the acquisitions of the navies of many nations.[16] A century later, Mahanian thought is used to draw parallels with space power. Traditional sea power notions such as 'choke points' have been transferred to space and the economic benefits of expansion into space are analysed extensively.[17]

Similarly, air power has been evoked in space power theory. After World War II, the US military predicted that space would become an important field of military advancements and profoundly transform military operations. US Marine Corps Major Franz Gayl argued that '[a]s with aviation, access and technology will drive forward to exploit any and all warfighting relevance, application, and advantage from space, quite independent of a nation's will to prevent it'.[18] If previously air dominance capabilities were key to global control, the control of space could mean potential future mastery of the world.[19]

Yet, while the expansion of human activity to an unexplored and ungoverned new area does lend itself to a certain analogy, many factors hamper such comparisons. Comparing space flight to air flight, James Oberg, one of the prominent recent theoreticians of space power, points out that 'there have been no warriors in space; there have been no weapons fired from space against terrestrial targets; and there have been no space-to-space engagements".[20] Furthermore, '[o]ther than the commands to keep a craft in a desirable orbit, there exists little other control over space assets by US military space organizations'.[21] Finally, '[to] a much greater degree than any other current human activity (especially those associated with "air power" or "sea power"), space flight is disproportionately vulnerable to breakdowns – accidental or deliberate – at these choke points'.[22] It is beyond the scope of this book to discuss air and sea analogies more comprehensively. It is sufficient to say here that although analogies are certainly helpful, the unique nature of space requires a unique conceptualization of space power.

Based on the above and other considerations, different definitions, emphasizing different aspects, of space power have been provided by various scholars. One of the best-known, rather general working definitions was offered by the 1998 US Air Force Doctrine Document, which defines space power as 'the capability to employ space forces to achieve national security objectives'.[23] In much the same vein, a 1998 Rand study defines space power as the 'pursuit of national objectives through the medium of space and the use of space capabilities'.[24] Other definitions focus exclusively on national security and military aspects, such as the more recent definition proposed by Professor John M. Lodgson, stating that 'space power can be defined as using the space medium and assets located in space to enhance and project US military power'.[25] Oberg adopts a rather wide definition arguing that '[s]pace power is the combination of technology, demographic, economic, industrial, military, national will, and other factors that contribute to the coercive and persuasive ability of a country to politically influence the actions of other states and other kinds of players, or to otherwise achieve national goals through space activity'.[26] Beyond these attempts to define space power, no comprehensive space theory has emerged so far.

Although today there is no single unified theory of space power, there is no doubt that space power has played an important role in recent history. As Captain John Shawn argues in his essay *The Influence of Space Power upon History* (1944–98), the nature of space power is fundamentally changing: while strategic space technologies used to be employed for predominantly military and national security uses, today space power influences a wide array of national areas, including civil, social and military ones.[27] Writing at the end of the 1990s, he rightly predicted that the influence of space power would increase continuously. In the post-Cold War era, the non-military applications of space technologies have significantly increased in performance. Space power today goes beyond traditional military, security and economic applications and comprises many potential opportunities in civilian and commercial domains.

I would therefore argue that any conceptualization of space power needs to take into account its multidimensional nature. In other words, space power not only includes all of the hard power tools, usually referred to as military and economic might, but also soft power tools, that is, the ability of a country to 'obtain the outcomes it wants in world politics because other countries want to follow it, admiring its values, emulating its example, aspiring to its level of prosperity and openness'.[28]

Several theorists have previously highlighted the many dimensions of state power. For example, at the beginning of the twentieth century, Mahan established a list of six factors that influence a nation's sea power: geographical position, physical conformation, extent of territory, size of the population, character of the people and character of the government.[29] More recently, it has been argued that similar considerations are applicable to space, and that space theory should draw on these and similar factors. For instance, USAF Lieutenant Colonel Martin E. B. France argues that at least three of these dimensions – geography and the character of the population and of the government – play a major role as regards space power.[30] Geographical position is crucial for the launch of a satellite because proximity to the equator allows the satellite to be boosted eastwards, and a large area of uninhabited territory around the launch site ensures that the population is not affected by accidents, debris, and so on.[31] The character of a people, and of a government, he argues, matters in the sense that a commercially oriented population will seek to exploit new possibilities and embrace technological progress. The proof of this, according to France, is that 'the two largest market economies in the world are also the two most robust space powers',[32] referring to the European Union (EU) and the United States. While there is some truth in France's arguments, such classical geopolitical concepts are by no means exhaustive. A view of space power that draws exclusively on classical geopolitical factors, with a heavy focus on military issues, would provide a biased and limited understanding.

Similarly, Oberg proposes a list of elements that make up the concept of space power, ranging from what he calls 'obvious hardware', such as space vehicles, launch sites and control sites, to the 'often overlooked human element', related to skilled human resources and a supportive society.[33] More specifically, he refers to hardware products such as satellites, laboratories and the manufacturing facilities required to produce space assets. He stresses the importance of a strong economy that will facilitate the funding of both governmental and private space programmes, a tradition of science and the intellectual climate in a country, the need for a pool of highly skilled engineers and scientists and therefore access to universities and research programmes.[34] Nicolas Peter, from the European Space Policy Institute (ESPI), joins Oberg in emphasizing that '[t]he foundations of space power range from obvious hardware elements (such as launch sites, launch vehicles, telemetry tracking and communications sites, on-orbit satellites and other spacecraft) to socio-economic elements (such as human capital) through to political and regulatory elements such as the number of seats in international organizations and other relevant bodies'.[35]

These approaches offer a more inclusive definition of space power by considering additional factors such as education, economics and internal politics. Yet, they still fail to provide a comprehensive *spatial* framework. Just as terrestrial prerequisites have consequences for a country's space programme, space power has, in turn, a significant impact on international politics on Earth. Mahan, Oberg, Dolman and other theorists tend to focus on the effects of the terrestrial preconditions of a country's capacity to access space and strive to analyse the resulting power configurations in space, usually as a mirror of geopolitics on Earth. However, the relation between space power and the terrestrial dimension is more complex and the simple transfer of our traditional 'beliefs about Earth-bound interactions into our thinking about the behaviour of states in space'[36] provides a biased picture of existing and future dynamics in space.

As Peter correctly observed in his work, it is important to distinguish between the prerequisites for a strong space programme and the effects of such a programme. The impact of space power on Earth is as important as any changes occurring in space, yet it does not get the attention it deserves in the current debate about space and global security. In fact, space applications have enormous implications for every aspect of our daily lives. Space power can help states to increase their capacities in many crucial areas, for example, to achieve military, technological, economic, educational or national pride goals. All these areas link back to the seven capacities of *meta*-geopolitics and state power introduced above. These seven dimensions not only determine a state's capacities *in* space, but also directly affect each of the capacities on Earth, thereby influencing the geopolitical structure.

In the light of these considerations, I would argue that a definition of a state's space power needs to comply with three core criteria. First, it has to consider the unique nature of space rather than recycling notions used for air, land or sea power. Second, it has to be broad enough to encompass dimensions that go beyond military and other traditional uses of space. Third, it needs to address the effects that space power may have on power relations on Earth. In his work on Europe and space power, Peter has proposed a multidimensional approach to the latter that partly meets these three criteria. He defines state space power as the 'total strength and ability of a State to conduct and influence activities to, in, through and from space to achieve its goals and objectives (security, economic and political) to affect desired outcomes in the presence of other actors in the world stage and if necessary to change the behavior of others by exploiting the space systems and associated ground-infrastructure as well as political leverage it has garnered'.[37]

I define *space power* as

> the ability of a state to use space to sustain and enhance its seven state capacities as outlined in the *meta*-geopolitics framework, namely (social and health, domestic politics, economics, environment, science and human potential, military and security, and international diplomacy). In addition, the governance and sustainability of space power will need to employ a '*symbiotic realism*' approach to global relations and a '*multi-sum security principle*' approach to global security. Ultimately, space will either be safe for everyone or for no one.

In comparison with the existing approaches, the definition of space power proposed above takes space as the point of departure of the analysis and investigates the way 'Earthly units'[38] utilize this new terrain to enhance their geopolitical posture on Earth. Moreover, this definition goes beyond the analysis of '*power enhancement*' through space to address the critical issue of '*power sustainability*' in a globalized, competitive and interconnected world. It draws on the postulates of the '*symbiotic realism theory*' *and* '*multi-sum security principle*' to advocate non-conflictual competition and absolute rather than relative gains in space. The section that follows explains in detail each of the seven capacities of state power outlined by *meta*-geopolitics and lays the theoretical foundations of two space governance paradigms: '*symbiotic realism theory*' and '*multi-sum security principle*'.

Seven dimensions of *meta*-geopolitics and state space power

Social and health issues

The first state capacity of *meta*-geopolitics, social and health issues, assesses three intertwined and interrelated dimensions: demography, social cohesion and health. First, *demographic* problems such as extreme population growth or decline can influence the geopolitical strength of a country, just as high numbers of educated people or high levels of unemployment can. In some countries, rapid growth of the young workforce, which cannot be absorbed by the jobs market, can lead to social unrest and political instability. Other countries, by contrast, are confronted with a shrinking number of young people and an aging population, implying less productivity and higher health costs and public spending on retirement, thereby affecting the economic performance and geopolitical position of a country. Moreover, in classical geopolitics, the size of the population is an important factor in geopolitical strength. Smaller countries do not have the same strength in regional talks, and have had to find other solutions to gain influence and power in the international arena.

Second, *social cohesion*, defined as 'the absence of large social and economic divisions within society',[39] can be affected by internal inequalities and divisions, such as wide economic disparities between rich and poor or ethnic divisions in the population. Societies with a higher level of social cohesion tend to be more resilient in times of crisis and adversity and often have more stable government structures.

Third, *health* issues have a serious impact on a country's capacities, and public health issues can have strong implications for its geopolitical status. Unhealthy populations have a huge impact on the economy, human capital formation and the general morale of the people. Problems in Africa and Russia concerning the spread of HIV/AIDS are an example of this. Improving social- and health-related outcomes will be a major challenge for countries in the future as recent years have witnessed potential and real pandemics. In sum, a population which is healthy and lives under stable social conditions is more likely to assert influence at the international level than a population afflicted by disease, unemployment and division. Conflict between ethnic groups, an aging population or a high prevalence of diseases can cause unstable conditions, reducing the relative strength of the country.

Space technologies and applications have strong potential to provide solutions and opportunities for life-enhancing improvements in social and health matters. Of the three areas outlined above, the strongest and most direct impacts of space applications concern national and international health care. Satellites have established themselves as an accepted medium for the remote delivery or the communication of medical services, as a tool for disease monitoring and as a means for training medical staff from a distance.[40]

Telemedicine, for example, is essential in critical situations in remote areas or simply where there is a lack of expertise and knowledge. An inexpensive method for providing consultative, diagnostic and treatment services, telemedicine has the potential to provide rural, remote and impoverished areas with medical care.[41] Telemedicine can be summarized as 'the use of medical information exchanged from one site to another via electronic communications for the health and education of patients or health care providers, and for the purpose of improving patient care'.[42] Telemedicine includes measures of tele-education, enabling distance learning and e-learning by health workers, it can particularly be highly beneficial for health education and training in areas that would otherwise not have access to such services.[43] The countries with the largest telemedicine programmes in the world are the United States, Canada, Japan, Australia, the United Kingdom, Norway and Sweden. However, research into and the piloting of such telemedicine projects and technologies have been carried out with success in countries such as Ghana and Nepal, where people live in rural areas with a lack of hospitals and specialists. This can be a way to address health issues, and if telemedicine is incorporated efficiently into the health system, it can significantly improve health care in those countries.

Satellites can also help monitor and carry out surveillance of the spread of an infectious disease. For example, they are used with high precision in identifying malaria-free areas. Moreover, satellite technology can save lives by providing early warning of extreme cold or of heatwaves. Such a case occurred in 2007 in north-eastern Kenya. Thanks to data collected by NASA satellites, a team of US researchers from NASA's Goddard Space Flight Centre noticed that temperatures in specific parts of the Indian Ocean were 0.5 degrees higher than usual. The same phenomenon had been observed in 1997, resulting in higher temperatures, increased rainfall and an abundance of mosquitoes. Combined, these conditions had led to one of the biggest outbreaks of Rift Valley Fever in East Africa. However, this time, thanks to satellite data, the Goddard researchers recognized the similarities between the two situations and constructed models of disease spread before they occurred. This allowed them to warn the Kenyan Government, which, in turn, was able to inform the population, distribute mosquito nets and take other preventative measures before the outbreak of the disease.[44] Information collected by satellites therefore prevented a major outbreak of Rift Valley Fever, saving many lives and protecting economic and human resources.

The impacts of space applications on demographics and social cohesion, while indirect, are no less relevant. Public health improvements can prevent social unrest and contribute to a more productive workforce. Satellites can increase the well-being of remote populations, but also connect these areas to central authorities, resources and technologies. This means that populations in remote areas can be better integrated into the country's activities.

Thus, space-based applications have a huge potential to contribute to the improvement of health and social conditions worldwide and even minor improvements in a country's general level of health can result in an overall improvement in its geopolitical strength. Yet, as with many issues, technology and the resources offering such benefits in health and social areas are more available in developed countries, and could therefore create asymmetrical improvements in health care and social conditions, which might widen the gap between developed and developing countries. It is the traditional space powers, Russia, the United States and the EU, that are leading the development of space technologies, but developing countries are in dire need of such tools. Development of their space assets and expertise in areas of observation, communication satellites and other associated technologies are essential to improving preparedness levels, surveillance and the efficiency of early warning. For developing countries, improvements in health care through satellites and other space applications could have a huge impact, and make a significant contribution to improving the welfare and health levels of the population.

Domestic politics

The domestic political situation in a country has a serious impact on its geopolitical strength. A stable domestic political situation will increase a state's ability to act in foreign affairs, just as an unstable regime will have problems legitimizing its actions with its population. Unstable domestic conditions also decrease the chances of a country being able to efficiently use its strategic location or strategic resources in order to enhance its geopolitical strength.[45]

Domestic politics are the result of the interaction of many different actors. In addition to governmental actors, non-state actors are increasing their role in domestic politics through lobbying groups, religious groups, the media, the Internet and so on. Companies and organizations are also gaining increased influence over domestic politics, including for example US space companies, the satellite industry, the telecommunications industry as well as the military and defence industries. They are pushing agendas which significantly affect policies and domestic priorities.

National prestige is an important part of a successful space programme. The desire to increase status in the international arena has led to an expansion of national space capabilities in many countries. The prestige that a successful space programme can gain is a tremendous unifying factor in domestic politics and increases the influence of relations with competing nations. The space race between the United States and the Soviet Union for instance was strongly connected to national pride and domestic political incentives. Iran is another example of a country where national pride over its space programme is encouraged and celebrated. North Korea also uses satellite launches as a morale booster for its population. Space technology is a highly advanced science and space launch nations are at the forefront of the world's technology race. Giving the domestic population a sense of national pride is connected to a successful domestic space programme.

On the downside, domestic politics can also limit space programmes by cutting budgets, reprioritizing policies and changing governments. Domestic politics are also likely to put pressure on research and development (R&D) to produce useful outcomes that will have clear benefits. It is difficult for institutions such as the US Congress to justify funding research programmes or allocating more money to them when results might not be forthcoming for another 50 years. The development of space capabilities is a long-term project, which extends far beyond the normal election period. In democracies, politicians and officials have short-term time frames to produce results from their political decisions. The recent worldwide financial crisis has also put a lot of pressure on government budgets, and space programmes could receive less funding as a consequence.

Space technologies can play an important role in countries with unstable domestic conditions, affected by civil war, the existence of rebel forces and breakaway attempts by certain regions. Satellites are key assets for such domestic challenges – they can be used for monitoring tasks such as border control and surveillance of strategic locations. The extremely high-resolution capability of today's satellites can make this technology a useful tool for preventing crime, stabilizing insecure areas and monitoring borders and conflict zones.

Economics

It is sometimes argued that economic power has become more important than military power, and that economic competition has taken over as the main competition between states.[46] While this is not completely the case, economic power is certainly of tremendous importance to international geopolitics. In recent years there has been an increase in the power of economic players and of states with strong economies. Natural resources may have been the most useful resource in the past, but trade and transnational commerce have increased the importance of production and economic stability at the international level.

The geopolitical implications of economics are affected by natural resources, access to international markets, energy reliance, technological innovation, strong domestic markets, and access to and influence over important financial institutions, such as the International Monetary Fund (IMF), the World Bank and the World Trade Organization (WTO). An economically strong country cannot be ignored by the international community and will be a significant player in world politics. The recent economic crisis has demonstrated the importance that world leaders place on functional economic markets, and recovery and crisis management have been at the top of the agendas of recent G-8 and G-20 meetings. Strong economic ties between countries also facilitate cooperation in other areas and can make a group of smaller states more powerful.

The EU developed through economic cooperation, enhancing the power of its member states at the international level. Economic growth in Brazil, Russia, India, China and South Africa, the so-called BRICS, has made these countries key players in the international arena. Economic strength can also increase a country's influence in international organizations such as the IMF and the World Bank.

The use of space has a tremendous impact on a country's economy. The extensive collaboration between agencies, governments, commercial companies and networks of space actors, resulting in a sort of *economics of space*, has enormous potential. Proponents of space flight argue that the economic benefits of space technologies and assets occur through the creation of jobs and markets. The international space industry was one of the few industries in 2010 that despite the troubled financial climate managed to increase its commercial revenue – by 7.7 per cent above 2009 levels.[47] The space industry has grown into an important part of many national economies and employs a large number of highly educated workers. More indirect benefits include the fostering of innovation and technological spillover effects, from which a large number of sectors benefit, such as agriculture, communications, mining, space science and exploration.

It is clear that technology and innovation are key driving forces in the economies of today's globalized world. Space programmes and space applications are vital industries for both developing and developed countries, and it is also obvious that the beneficial economic impacts of space use have significant consequences for the geopolitical status of a country. One of the most prominent examples of the effects of space applications on the economic power of a state is the US NAVSTAR Global Positioning System (GPS). Originally created as a military navigation tool, GPS has become a commercialized system that is widely accepted and used around the globe. In addition to providing an economic stimulus to the US firms involved, GPS marks US technological dominance and therefore has significant implications for its geopolitical strength. Current attempts by the EU to establish an independent and competing global positing system, Galileo, are therefore motivated by a mix of geopolitical and economic reasons. Once completed, Galileo could challenge the US monopoly in this market. This will give significant economic advantages to EU member states and the other actors involved, and reduce the worldwide dependence on US technology and thus US influence across the world. Similarly, Russia is taking steps to improve its global navigation satellite system, GLONASS, with a new generation of satellites. China is also improving its global navigation satellite network, working on a second generation system, Compass.[48]

The promise of emerging space-related markets is evident from the recent surge in the number of private and commercial actors in space. Private access to space is increasing exponentially, and so is its commercial exploitation. Remote sensing and mobile satellite services are only two examples of significant financial opportunities. With the shift of some government space services towards privately owned companies, the roles of the public and the private sector are being redefined. Well aware of such trends, the US Government puts a strong emphasis on private-public partnerships and seeks to externalize some traditionally state-run activities to the private sector, such as putting NASA astronauts into orbit.[49]

Finally, there are some emerging markets in the space economy that could have a huge effect on strategic and geopolitical situations. The more countries access space for economic benefits, the stronger their economies become. The cheaper space assets become, the more players will take part in space activities and increase the global market for space-related products. A stronger market will also increase the capabilities we have in space, perhaps leading to important advances that are not even dreamed of today.

The environment

Recent developments in climate change, together with worries about the decline in natural resources such as water, oil and raw materials have made the environment a stronger than ever aspect of geopolitics. Environmental threats are one of the most salient current security problems, and a state's environmental conditions will affect its geopolitical position. Water and air pollution can have a negative impact on population health, and environmental conditions will play an important role in future geopolitics. States with guaranteed access to adequate supplies of water, for example, could improve their geopolitical position in the international system in situations of water scarcity. Global warming and its effects on the environment all over the world could also drastically change the geopolitical situation. For example, flooding can lead to large movements of people, the melting of the ice caps will create new incentives for states around the Arctic to change policies and environmental disasters, such as hurricanes or earthquakes, will place massive burdens on the infrastructure and economy of countries. Natural disasters can cause serious harm to a state's strength, and its ability to defend itself or act fully in the international arena. Environmental disasters rarely respect national borders, making them international issues that demand coordination and burden-sharing between states and regions. Such coordination and cooperation is not always easy and the results could either enhance the strength of a country or diminish it. Energy and environmental security are not just the problems of a single state, but major critical challenges for the entire world.[50]

The environment and space exploration have been closely connected since the dawn of the space age. Space applications developed and made accessible by NASA and others, in particular environmental satellites, can play a huge part in avoiding or mitigating environmental problems, and can provide a better understanding of our world and our impact on Earth. Space technologies such as remote sensing satellites can provide early warning, assist urban and rural planning and show changes in coastlines, forest densities and ice caps. Weather satellite technology, one of the first satellite technologies, developed in the 1960s, has had tremendous impacts on weather forecasting and hurricane prediction, and the subsequent protection of the populations concerned.[51] Earth monitoring satellites first emerged in the 1970s and provide information about land and sea.[52] They can help to identify high-risk areas and monitor the levels of gases in the atmosphere. Like most advanced technologies, they are mostly used by developed countries, but new and cheaper satellites, such as micro-satellites, could potentially assist developing countries to combat environmental problems. In countries such as China, where extensive economic expansion has led to increased pollution, space technologies can mitigate ongoing problems and prevent them from hampering economic growth.

Both NASA and the European Space Agency (ESA) are involved in different climate change projects, and emerging countries such as South Africa, India and Brazil use their satellite resources for extensive environmental monitoring. Environmental problems and climate change will affect all countries. An advanced space programme can increase a country's potential to deal with such issues, giving it a relative advantage over other countries.

Similarly, energy scarcity is one of the most pressing environmental problems today. Space technology is developing rapidly and a number of techniques could have a huge impact on a society's energy supply in the future, such as the exploitation of celestial bodies. Studies have confirmed that there is a wide range of exploitable materials in asteroids, such as iron, platinum, water and trapped or frozen gases, and that the use of space-based asteroid mining could be a critical issue in future geopolitics.[53] Another option would be the exploitation of space objects approaching the Earth, called Near Earth Objects (NEOs).[54] Although this is not yet possible, a country with full access to these technologies would undoubtedly increase its geopolitical strength in the future. Space-based solar power may be crucial in the future. If one state could make full use of the potential of space-based solar power, it could become completely independent of other states with regard to its energy supply. Few countries today have that luxury. Energy, and especially oil, plays a central part in security-related debates. It is crucial for states to have sufficient sources of energy and control of a new and important source of energy could be the key to increased power.

Science and human potential

A country's scientific know-how and the skills of its population are crucial indicators of that state's power, and of its ability to project such power abroad.[55] Investment in R&D, high-quality education systems and the resulting intellectual capital and skills are indispensable for international influence and power.

Technology drives change and progress, and a country's level of power can change drastically in relation to its technological innovation capabilities. While technologies and science might not be obvious tools of power in themselves, they can work as multipliers, enhancing current capabilities and developing new ones. There have been numerous occasions in which a crucial new technology has drastically changed the world. The development of the nuclear bomb is one of the best examples. Communications technology and the Internet have made the world more accessible and have had huge implications for power in a globalized world. New technologies can enhance a state's economic capabilities as well as its military capacities. Development of a higher level of science and technologies will also have spin-off effects for a country's education level and improve the strengths and skills of the population. A highly educated population will foster economic growth and engage in a virtuous circle to further enhance technology and science. The level of human resources within a state is also an important factor in measuring its geopolitical strength. A steady supply of new members of a qualified workforce is a prerequisite for economic growth and domestic stability. A country's human potential and technological level can be used to advantage when a state lacks other important components of geopolitics, such as a large population or a large land area. By investing in both general education and more specialized R&D, a county can become more competitive over time. By having high levels of knowledge and technology, a country can also achieve strategic monopolies in certain industries, leading to increased influence in world politics.

The space industry has a clear impact on a country's scientific level and its human potential. It is a strategic and high-technology industry with enormous spillover effects on other sectors. A successful space programme is a clear sign of a technically advanced nation. The space race between the two superpowers during the Cold War showed the significance and influence of advanced scientific and technological capabilities. Both the United States and the Soviet Union invested heavily in R&D and scientific studies. These efforts gave them dominant positions in space, positions that were not challenged until 20 years after the end of the Cold War. Having a monopoly on a specific technology can offer significant geopolitical advantages. The US-owned GPS, in addition to the economic benefits it brings, is an excellent example of a space technology providing such geopolitical opportunities. By controlling the world's need for navigation systems, the United States gained the significant advantage of being able to control other countries' access to and use of this technology. The GPS can be shut down or restricted by the United States in times of conflict. In the words of Gulten Acar, 'historical breakthroughs in technology in civilian and military fields, as much as they advance the common aspirations of mankind, are closely linked to geopolitical maneuvering and interests. Unless Galileo and GLONASS [and Compass] become fully functional, the only global positioning system will remain under the control of the US, which means that strides in strategic technology is also part and parcel of global political positioning'.[56] For aspiring geopolitical powers, this confirms the need to continue to invest in the sciences and technology in order to maintain a technological advantage.

The geopolitical implications of US predominance in space technology are also visible in the context of international arms control, where China and Russia are advocating a ban on space weapons in order to reduce the asymmetric technology levels in space by relatively increasing their influence. The development of space technologies by other states (India, Iran, Israel, North Korea and Pakistan) in critical geopolitical areas shows that technological know-how has many advantages and can be used to increase the projection of power. Furthermore, most space technologies entail dual-use capabilities, meaning that the development of civilian space applications can also serve military purposes. For example, India, Iran and North Korea have managed to develop the indigenous technology required to launch a space rocket, which in turn can be used to create missiles capable of carrying nuclear weapons. Increased technological capabilities in space have a clear effect on the geopolitical situation on Earth, and the close relationship between technological development and increased military capabilities makes space technology a powerful tool for states seeking to enhance their position in the global community. Advanced technical states can also sign comprehensive cooperation agreements with less developed countries that are rich in resources. Space cooperation between China and actors such as Brazil, Venezuela and Nigeria might result in advanced Chinese technology helping a country to provide enough natural resources for its economic development. Technology can be an efficient tool for enhancing such exchange agreements.

Space technologies and education are also increasing our awareness of ourselves, about the universe, and about the condition of our planet. They can generate significant spin-off effects, such as knowledge about environmental conditions and the threats we might face. They provide more evidence about our own existence and about the origins of our universe.

Military and security issues

Traditionally, the strength of a state's military power and its ability to tackle national security challenges have been used to gauge relative levels of state power in international relations. Increases in military capability can change the relative power balance between states, making this one of the core issues for addressing a country's security needs.

Nuclear weapons are only one example of an important military capability that has had serious power implications for its possessors, their allies and neighbouring countries. It is no coincidence that the five recognized nuclear weapons states are the same states that have a permanent seat on the United Nations Security Council. In sum, the level of technology, the percentage of the gross domestic product (GDP) spent on defence and the quality of a country's arsenal together contribute to the position of that country in the international community.

An increased capability in military affairs can come about through many of the other capabilities discussed in this chapter. For example, enhanced economic growth can result in increased investment in arms and the defence industry. International diplomacy can result in alliances, such as the North Atlantic Treaty Organization, where the nuclear might of some countries is used to defend other member states; or such organizations as the EU, where a joint foreign policy or common forces are discussed which would lead to significant increases in power for some of the smaller member states. Developments in technology and science can make a major contribution to the expansion of a country's military capabilities, through the development of more efficient weapons, unmanned weapons or new weapons and delivery systems. Technological advances in physics led to the creation of nuclear weapons, the weapon best known for changing geopolitical balances and providing countries with the most comprehensive defensive or offensive capabilities. New technology will also reduce the importance of geographic difficulties for a state, making it possible to fire weapons long distances across continents, over mountain ranges or through seas. Surveillance technologies will also increase knowledge about an opponent's forces, providing efficient targeting data which would be useful in case of an attack.

It is a widely accepted fact that space has to date been militarized but not weaponized.[57] Military uses of space have led to such enormous increases in capabilities that some believe that it is no longer possible to conduct warfare without access to space assets. Space applications and technologies support nearly all terrestrial air, sea and land force operations, and most military operations rely and depend heavily on them.[58] Many navigation, surveillance and communications satellites in orbit at the moment provide services for military purposes. These assets are not weapons in space, but they are force multipliers as they increase the efficiency of military operations on land, sea and air. Such space force enhancement has changed the geopolitical power of existing military capabilities. Satellites and navigation systems are now crucial for conducting military missions, and military dependence on space is increasing. In modern day warfare, space technologies will increase the ability of military leaders to observe an ongoing conflict, rapidly assess threats and potential opportunities, identify new targets and arrange attacks on such targets.

It is also worth remembering the historical links between satellite technologies and the nuclear forces of dominant world powers. During the Cold War, nuclear deterrence relied on satellites for early warning, communications and targeting information. For this reason, '[t]o interfere with the satellites of major powers has meant – and continues to mean – the possible use of nuclear weapons, since major powers could view attacks on satellites as precursors to attacks on their nuclear forces'.[59]

The potential weaponization of space thus presents a serious challenge to global security. The potential for the application of space technologies and space assets to the military and security-related sectors of a state is huge, and the militarization and weaponization of space has probably been the most discussed topic in the literature since the dawn of the space age. The long-standing debate between advocates of the peaceful use of space for the benefit of all humanity and advocates of military uses in space has yet to come to a conclusion. However, today, most world leaders seem to agree that space is 'a de facto sanctuary from war and weapons'.[60] The principle of the non-weaponization of outer space is affirmed in the 1967 Outer Space Treaty (OST), the leading international framework on the use of space, which asserts that activities in space should enhance the well-being of all countries and all humankind, prohibits any state from assuming sovereignty in space and bans the placement of nuclear weapons in space. Nonetheless, the growing number of space-faring actors has resulted in an increasingly complex and competitive geopolitical situation in which states' desires to protect their investments and security may ultimately lead them to introduce weapons into space and result in a vicious global space arms race.

Extensive utilization of space and particularly the military uses of space technology have thus the potential to create a global climate of insecurity by enhancing current conflicts and becoming the trigger for new ones. Beyond this well-known national security component with its strong military dimension, recent developments in space have the potential to affect global security in its human, environmental, transnational and transcultural dimensions as well. In my opinion, the traditional approach to security is no longer adequate to deal with the twenty-first century globalized international community and its rapid technological developments. Therefore, as is argued in one of my earlier works, *The Five Dimensions of Global Security: Proposal for a Multi-Sum Security Principle,*

> [i]n a globalized world, security can no longer be thought of as a zero-sum game involving states alone. Global security, instead, has five dimensions that include human, environmental, national, transnational, and transcultural security, and, therefore, global security and the security of any state or culture cannot be achieved without good governance at all levels that guarantees security through justice for all individuals, states and cultures.[61]

Each of these five dimensions refers to a different set of substrates. Human security refers to the security of individuals; environmental security refers to the security and survival of the biosphere; national security refers to the security of a state's population as a whole as well as to its economic position, its political system and its organizational system; transnational security refers to the security of states against transboundary threats and transcultural security refers to ensuring that a plurality of cultures and sub-cultures can coexist without causing security threats.[62]

Space has unique and remarkable characteristics, it is perceived as a common sphere going beyond national and cultural jurisdictions and a truly global common that must be used for the benefits of all of mankind. The emergence of advanced space technologies, coupled with the increasing number of actors in space, has created new security threats such as space debris, amplified the risk of weapons proliferation and increased dependence on satellite systems that directly affect the five dimensions of global security on Earth. As a consequence, unilateral actions in space encouraged by national security concerns cannot improve global security but only destabilize it. It is therefore essential for policymakers to address all five dimensions when discussing ways to secure space for future generations and to provide multilateral solutions in order to preserve our common heritage.

Looking at space through the '*multi-sum security*' lens and its five dimensions provides additional insight into how to promote security and sustainability in outer space – an area that has grown in importance for the whole world. Together, these five dimensions reveal a different reality, where space security is no longer a '*zero-sum game*' and unilateral measures are counterproductive and create even more insecurity for all.

According to the *'multi-sum security principle'*, a state that seeks to maximize its security needs to apply principles of justice in all five spheres, as injustice in one sphere will lead to insecurity in all of them.[63] The opportunity for mankind to use space for peaceful purposes, irrespective of which state or commercial entity, clearly shows that common security interests are prerequisites for national security interests. In other words, 'it is essential for both groups [military and the politicians] to recognize that the hope for victory is no longer valid, and that we must instead strive for common security, that is, security *together with*, and not *against*, a potential enemy'.[64] Safeguarding space while using space applications to enhance the just development of the seven power capacities of *meta*-geopolitics will create favourable and lasting conditions for sustainable use of space and ensure global security. The utility of this *'multi-sum security principle'* is discussed further in Chapters 4 and 6.

International diplomacy

International diplomacy, including the membership of and activity in international organizations, is an important tool for states to gain influence in global decision-making processes. Permanent membership of the United Nations Security Council is a significant position, and increased demands are being made by states to either eliminate its right of veto or expand the number of permanent seats. Gaining a permanent seat on the Security Council would significantly increase a state's power and credibility on the world stage. Influential states are thus more capable of putting their priorities on the global agenda and defending their national interests in such arenas as high-level EU, Security Council or G-8 and G-20 meetings. However, a state's diplomatic leverage is not always a direct result of its military or economic strength, or even its population size. Small countries can gain influence through diplomatic activities in regional, multinational or global organizations. For example, a country can have more influence over the outcome of certain negotiations if it is the host country of an important conference or if it presides over a significant meeting. It can also gain significant influence through a non-permanent seat on the Security Council. Smaller countries can also form temporary alliances around a core issue, and in this way manage to influence larger states more than would be possible alone.

As regards space, the influence of an actor in the global discussions on space matters and its visibility in the related global institutions constitute a measurement of a state's prestige and power. Enhanced diplomatic capacities can be used as a way of curbing rapid development that some countries cannot afford to keep pace with. Space technologies, however, are not only for prestige. They are also a resource for rapid and accurate information gathering. Since information is said to be a tool of power, space applications can have a significant influence on global negotiations and international confidence-building measures.

For example, satellites are a valuable tool in monitoring the implementation of, and verifying the compliance of world actors with international rules. For instance, observation satellites are increasingly used by international organizations and non-governmental organizations (NGOs) to monitor arms control and environmental treaties, cross-border disputes, forced migrations and any other abuses of humanitarian rules. Satellite technology has undeniably increased the scope and precision of existing verification instruments, but its utilization as a monitoring tool under multilateral treaties remains controversial. The major concern is that the pictures taken by remote sensing satellites can disclose strategic information about the national security of a state and thus infringe its national sovereignty and territorial integrity.[65] Despite these obstacles, remote sensing technology is used to monitor some multilateral disarmament and non-proliferation treaties. For example, states parties to the Treaty on the Non-Proliferation of Nuclear Weapons (NPT) are subject to International Atomic Energy Agency (IAEA) inspections, which use images from civilian and commercial satellites to verify the veracity and completeness of the declarations made by states and detect any suspect activity.[66] While satellite images alone are certainly not sufficient to ensure proper monitoring of such an important treaty as the NPT, they contribute to reinforcing the verification measures and thus increase confidence between states parties.

Remote sensing satellites are also used to monitor climate change agreements, such as the Kyoto Protocol to the United Nations Framework Convention on Climate Change (UNFCCC).[67] Implementation of the Kyoto Protocol is carried out in collaboration with the World Meteorological Organization through the Global Climate Observing System (GCOS).[68] Remote sensing technology represents an integral part of systematic observation of the treaty as it enables a close monitoring of greenhouse gases. Today, satellite technology is gaining importance in treaty monitoring and compliance verification. The images it provides can serve as confidence-building measures and, as a result, enhance trust and foster international cooperation in the fields of disarmament and climate change. A satellite-based verification regime must, however, carefully balance the preservation of states' strategic interests with the international community's need for robust confidence-building and verification measures.

Space agreements and institutionalized procedures for space conduct are also essential for preventing conflicts and ensuring the sustainability of the individual space powers. Given the growing economic value and strategic importance of space to state and non-state powers, a constantly increasing number of actors entered and are now competing in space. The countries with the most space assets and capabilities are also the countries that are the most vulnerable to intentional or unintentional threats in space. To sustain their power capacities the space-faring nations have to protect their critical space assets from irresponsible and harmful actions. The application to space of the '*symbiotic realism*' theory of international relations, developed in my earlier work entitled *Symbiotic Realism: A Theory of International Relations in an Instant and Interdependent World*, ensures the sustainability of national space powers through '*absolute gains*' and '*non-conflictual competition*' in a realist framework.[69] The '*symbiotic realism*' theory states that globalization, interdependence and instant connectivity characterize our modern societies. In this interconnected world 'the soft power capabilities have become increasingly significant components of a state's relative power, such as a state's ability to portray a favorable image of itself and its policies abroad'.[70] Moreover, '*symbiotic realism*' highlights that non-traditional actors, such as NGOs and multinational corporations are playing a more active role in international affairs and need to be taken into account when analysing contemporary international relations.[71] '*Symbiotic realism*' is an international relations framework that enables a comprehensive analysis of the international diplomacy mechanisms and is particularly relevant for the understanding of the role of the diplomacy in space.

Space is a *global common,* characterized by a high degree of interconnectivity and interdependence. Continuous advances in technology offer new applications and opportunities for global stakeholders to assert their power and thus improve their relative posture on the international stage. Political rivalry and innate competition make them seek further areas they can take control of. These developments render space actors increasingly reliant on space-based systems and particularly vulnerable to their disruption. According to the '*symbiotic realism*' framework, to sustain their space power in a highly competitive and anarchic environment, all stakeholders have a clear self-interest in keeping their own assets safe. In other words, the geopolitical opportunities for global actors to use space in order to improve and maintain their power on the international stage make it essential to build a sustainable security regime on the principles of common interests, cooperation and justice that enables non-conflictual competition. The relevance of '*symbiotic realism*' to space governance is further discussed in Chapter 6.

Conclusions

Since the dawn of the space age, space has had crucial implications for international power relations. However, with the end of the Cold War, the number of actors in space has increased and the use of space applications has broadened, with important geopolitical and strategic implications for the structure of international relations. Non-military and commercial space uses have significantly gained in importance and performance. They have the potential to influence many areas of our daily lives. As is discussed above, this has significant implications for the concept of space power. While previously it was mostly considered a state-centric, geography-bound and military domain, today it encompasses a much broader range of capacities and is used by state and non-state actors to enhance their geopolitical position on Earth.

Already fast-paced developments in space technologies and global actors' space capabilities are moving faster than ever. Space programmes appear to be a source of increased influence and geopolitical strength in the world, through their impact on enhancing human capacity and improved technological levels, advanced environmental monitoring, enhanced military capabilities, diplomatic efforts in international institutions, improved health applications and growing economic implications as well as increased national pride. These seven dimensions are interlinked and affect each other through constant overlaps. Consider for instance the rapidly growing private sector's space presence. From communication satellites to weather tracking devices to cable television equipment, outer space is full of civilian and privately owned space technologies. The presence of these private companies makes space much more accessible to countries that may not have their own independent space capabilities. However, the challenges presented by this privatization of space are numerous. States need to consider how a commercial presence in space might affect space-based military endeavours. Economic issues are concerned because civilian satellites could easily fall victim to a space battle, resulting in communications outages and major economic disruptions for those countries involved in the battle as well as others that are not. Diplomacy may be involved because as satellite orbit paths become ever more congested, states and the international community may have to reassess the criteria for launching commercial and government satellites and start regulating satellite numbers.

Finally, the many major advances in space-related technologies, such as precision navigation, data management and high-resolution remote sensing, have led to a global dual-use dilemma. Technologies that have valid, peaceful applications may also have military usages – a reality that makes it doubly difficult to establish guidelines for acceptable, legal technologies.

In the light of these examples, space provides us today with an incredible range of technologies and opportunities. Yet, in parallel, there is a growing appreciation of the urgent need for regulation. Space does not belong to anyone and only a system of global governance built on justice for all the parties involved can ensure that it is kept that way. Having outlined the theoretical basis of this book and clarified the concepts of '*space power*', '*multi-sum security principle*' and '*symbiotic realism*', Chapter 3 uses the framework of *meta*-geopolitics to look at the development of core space technologies such as communications and remote sensing satellites, global positioning systems and space launch rockets and their potential impact on the seven space power capacities of a state actor.

3
Space Technology and *Meta*-Geopolitics

The multipolar and less predictable world that emerged after the Cold War, shaped by globalization and events like 9/11 and its consequences, has seen space capabilities gain importance and become a core element of a state's strategic thinking.[1] Strategic, here, refers to means that a state needs to reach its goals, objectives and desired outcomes in each of the seven areas of power defined by *meta*-geopolitics. As applied to space, it means that domestic politics, the economic situation, demographic trends, health and social conditions, access to natural resources, vulnerability to natural disasters, the availability of a highly skilled workforce and technological know-how, national military and security considerations, as well as diplomatic leverage will together determine the importance a state gives to a national space programme. It will decide how large it will be, the strategic priorities it will have and whether it will take a confrontational approach to the other actors in space. A wealthy country which considers its military power to be an important part of its national security may be more likely to allocate resources for the establishment of a space launch capacity than a less developed country that believes in neutrality. Similarly, an actor that is frequently threatened by environmental disasters such as hurricanes would be more likely to invest in environmental monitoring satellites.

Given the importance of space to contemporary geopolitics, this chapter examines the development of space technologies through the prism of *meta*-geopolitics to assess how applications of space technologies might affect the seven power capacities of a state and, as a consequence, alter its international posture in comparison with other actors. In order to analyse recent developments in space, it is important to understand the scientific and technological underpinnings of space power.

Man cannot enter space without technology and any exploration or exploitation of space is dependent on highly complex and sophisticated technological equipment, scientific know-how and skills, all of which are evolving at an exponential rate.

The chapter starts by providing a brief historical overview of the evolution and growth of space technologies through three major phases that correspond to its military, civil and commercial development. However, as pointed out by Nicolas Peter, space power 'is not simply satellites and access to space, it is anything and everything a country can achieve through space'.[2] Therefore, this chapter focuses on the most prominent technologies currently in use, such as communications and remote-sensing satellites, global positioning systems and space launch rockets and examines how different strategic uses of these technologies might influence a state's space power capacities. More precisely, based on the *meta*-geopolitics framework and concrete examples, this chapter analyses the potential of space technologies to develop one or more of the seven capacities of statecraft and influence its overall geopolitical posture. The chapter concludes with a discussion of future space technologies and their implications for geopolitics and global security in the coming decades.

The development of space technology: A historical overview

Ever since the first man-made object was launched into space by the Soviet Union in 1957, space technology has been a symbol of power, knowledge and national prestige. Space technologies have given humans reason to consider the possibility of one day living on the Moon or on other planets and have inspired thoughts and ideas beyond the limits of existing technologies. However, space technologies did not grow out of a single unique invention or the emergence of specific know-how, space technology 'was the product of a convergence of various types of knowledge and know-how that already existed, and which, in some cases such as the laws of gravity, had done so for several centuries'.[3] Throughout the relatively short history of space exploration, the evolution of space technology has traditionally been divided into three different periods that correspond to its military, civil and commercial development. This section briefly outlines the rise, progress and maturation phases of space technology, discusses the interdependencies between the military, civil and commercial sectors and highlights some of the key technologies that have emerged out of this process.

The beginning of the space age

Outer space has always been fascinating to mankind, but the first space technologies leading to the physical exploration of space were the result of the struggle of two superpowers for supremacy in the 1960s and 1970s. Driven by Cold War competition, the United States and the Soviet Union competed with each other for control of the heavens in what is commonly called the *space race*. The space race was an important part of the ideological, cultural and, of course, military rivalry between the two blocs and contributed hugely to national prestige. More importantly, this competition resulted in many advanced and sophisticated technological developments.[4] The event that marked the dawn of the space age and the space race was the launch by the Soviet Union of the world's first satellite, Sputnik, '*travelling companion*' in Russian, in 1957.[5] In response, and driven by a sense of urgency to rescue US national prestige, the United States created the National Aeronautics and Space Agency (NASA) and launched Explorer I, the first US satellite, in 1958.[6] In order to demonstrate US technological and scientific superiority, in 1961 President John F. Kennedy proposed landing a man on the Moon. The crew of Apollo 11 reached the Moon in 1969, and Neil A. Armstrong and Edwin E. Aldrin Jr. were the first men to walk on the Moon – completing one of the most complex and challenging technological undertakings of the twentieth century.[7] Since then, the pace of technological advances has quickened, as has the worldwide spread of related information.[8]

Hence, much of today's space technology was developed during the Cold War, when technological advances were driven mainly by national security and military considerations. As such, space research and activities were primarily government-funded and organized by the military. It was during this time that the first satellites were launched in order to gather intelligence, although the list of potential satellite applications quickly expanded. Throughout the Cold War, the United States and the Soviet Union regularly deployed reconnaissance, early warning, intelligence-gathering and communications satellites.[9] The collection of information about the enemy's capabilities was crucial for the superpowers and was kept largely secret.[10] To facilitate strategic planning and targeting, the US military also operated several weather satellites and missile and early warning systems.[11] Satellites were also useful as a national technical means (NTM), a tool for the verification of arms control treaties.[12] There was an ongoing programme of the modernization and production of intercontinental ballistic missiles by the Soviet Union, as well as efforts to build anti-satellite weaponry and to create a reusable space vehicle, the space shuttle, by the United States.[13]

However, even though these early technological developments served national security purposes, the United States and the Soviet Union adopted a non-military approach to the space race. Great efforts were made to demilitarize satellite launches and the Moon landing and to make the competition look peaceful, thereby steering the public perception of space as a sanctuary.[14] This was necessary in order to legitimize the power of communism or capitalism as technologically superior, and demonstrate that the gains were as a result of the natural technological superiority of the winning ideology rather than a victory for military force.[15]

With the demise of the Soviet Union and the end of the Cold War, a new era for the military use of space applications began, but military dependence on space-based assets continued to increase. The 1991 Gulf War is widely considered the first *space war*, although, as General Kutyna pointed out, it was really 'the first space applications war'.[16] More than any previous conflict, military operations relied heavily on space technologies for the collection of weather data, for missile warning data and signals and for navigation. The United Nations Security Council condemned Iraq's invasion of Kuwait in August 1991 and, following the adoption of resolutions 660 and 678, the United States led a coalition of states in the largest military operation since the Vietnam War.[17] Satellites and other space-based assets had been used in previous conflicts, but they played an essential role in the Gulf War by providing coalition forces with advanced communication systems; a global positioning system; surveillance through optical, radar and infrared measurements and electronic information gathering. While it is commonly acknowledged that space technology was not the determining factor in the victory, it was nonetheless a significant development and a clear sign of the growing military potential of such systems.[18]

Using today's space technology, the military can obtain images, navigate, receive intelligence signals, operate communications systems and early warning systems and obtain accurate and advanced weather forecasts.[19] It is important to note, however, that the military applications of space assets are force multipliers, which increase the efficiency and accuracy of existing terrestrial weapons and other military means. Force multipliers can improve situational awareness, enable attacks to be carried out with greater precision and over a greater range, and improve the efficiency of support forces, as well as enhancing coordination between allied and friendly units. This can lead to significant changes and force improvements – an attacking force can be smaller and lighter, and operate in a more streamlined manner. These tools can also improve the speed and the accuracy of military units, as well as the ability to adapt to changing circumstances and unexpected events.[20] Although the first developments of space technology were driven by military considerations, the civilian use of space technology was important from the very beginning, and, since the end of the Cold War, has continued to increase in importance.

The rise of the civilian use of outer space

Space technologies initially developed for military purposes have been adapted for use in a variety of civilian applications. For example, space transportation, satellite communications, global positioning systems and remote sensing have contributed significantly to improving daily life on Earth. The civilian sector mostly concentrates on technology development, scientific innovation, space exploration and operational capacities.[21] However, the domains of civilian and military space use have been intertwined and mutually dependent since the beginning of the space age, and the mutual interdependence of civilian and military technologies has become even more accentuated in recent years.[22]

Indeed, most civilian space applications grew out of military objectives. The space launcher, for instance, which remains the only way to enter space and to place objects in orbit, was originally developed to enable the launch of intercontinental ballistic missiles (ICBMs).[23] In reality, the boundaries of the so-called space sanctuary are fuzzy at best as space technology has, by definition, dual-use functionality. Who is to say, for example, that a weather satellite could not be used deliberately to crash into a military satellite belonging to a rival? This would certainly make the weather satellite a weapon, but it is hard to imagine a world in which countries are banned from putting weather satellites in space. According to Joan Johnson-Freese, 'there is no distinction between space technology for civil or military use, since 95 per cent of space technology is dual use, and further – and really problematic – there is often little or no distinction between military technology that is offensive or defensive in nature'.[24]

Because of the inherent dual-use functionality of space technology and its multiple applications, it is no longer possible to categorize space applications as strictly civilian or military.[25] For example, commercially available remote-sensing technology may have had government funding and military origins in the past, even if it is now privately owned. Different technologies can also change their functions over time: commercial high-resolution systems or space imaging represent a civilian variation of earlier US military systems, and the US NAVSTAR GPS was initially developed by the US Air Force for military purposes but is now critical to monitoring climate change and understanding the environment, and it also has enormous commercial value in the global space services market.[26] Earth-observation satellites are an example of the many types of satellites that can perform both civilian and military functions, such as weather monitoring, reconnaissance or missile warning and defence.[27] The US Department of Defense (DoD) Meteorological Satellite Program gathers data from its own military weather services but, at the same time, receives data from the civilian Geostationary Operational Environmental Satellite (GOES) and National Oceanic and Atmospheric Administration (NOAA) satellites.[28]

Sputnik was the start of what has become a fast-moving adventure. States have become increasingly dependent on space-based systems to ensure their national security and a wide range of services to the population. The development of space technology has had an enormous impact on life on Earth and on human development. For a long time, civilian space activities were dominated by governmental actors. However, the high cost of space programmes coupled with increasing international competition opened up space to private actors and initiated its commercialization.

The commercialization of outer space

In addition to the many states with active space programmes, the number of private actors entering the domain of space is also increasing. Space has been opened up for commercial use, the best known uses being satellite communications and navigation systems, and satellite television and radio. Civilian communications satellites were the pioneer private sector activity in space. Although government agencies were the first to develop satellite communications, the situation changed in the early 1960s when companies such as Comsat in the United States and Intelsat in Europe introduced satellite technology to commercial telephony.[29] In line with the worldwide trends towards deregulation, privatization and liberalization that have shaped the global economy since the late 1970s, trust in the private sector's efficiency in the space industry has also grown.

In recent decades, the commercial use of space technology has increased rapidly along with the emergence of a huge space services market. High levels of competition among firms mean that the cost of space activities, such as launching commercial satellites, for example, has dramatically decreased and these economic benefits are driving further investment and innovation in space technology.[30] Moreover, increased private sector demand for new forms of networking, data management and digital applications has fuelled this boom and pushed satellite service providers to invest in new forms of technological combinations, such as communications and positioning satellites.[31] Beyond the media and telecommunications industries, space products and services have many other commercial applications, including health care, financial services and transportation.[32] Space applications are used today to improve telemedicine services, Internet connectivity and mobile satellite radio services, and for tourism and travel planning, navigation, resource management and entertainment.[33] An area of commercial space development which remains marginal is space travel for private citizens. Russia has been most prominent in this field, sending several private citizens into space. Commercially viable space travel has yet to become more widely available, although many private investors continue to study ways to make it feasible.[34]

It is interesting to highlight here the relationship that exists between the commercial space industry and civilian and military government agencies.[35] The US military, for example, increasingly relies on civilian and commercial space services and assets, such as 'integrated tactical warning and attack assessment (ITW&AA), weather and environmental monitoring, satellite communications (satcom), surveillance and reconnaissance, navigation and positioning, space control, ballistic missile defense (BMD), and force application (i.e., using weapons that travel through or are based in space)'.[36] We also witness the emergence of public-private partnerships in the field of space and this trend perpetuates and reinforces the existing interdependence between the space industry, governments and space agencies.

From satellite television to global positioning devices and data transfer, space-related technology is profoundly integrated into our daily lives. Space research has led to countless scientific breakthroughs and innovations that benefit additional domains such as health, the environment and education, thereby serving the global community.[37] Our increased use of space technologies has demonstrated the importance of the space industry in the economic sector. Commercial space creates new markets and plays a significant role in fostering economic development and growth. Beyond its numerous everyday civil applications, space technology continues to support a variety of security and military operations. However, as is noted above, the traditional boundaries between the military, civilian and commercial uses of space have become more blurred.[38] Many space technologies are dual-use in nature. The section that follows discusses the possible strategic applications of some of the most prominent space technologies and evaluates the potential consequences of such applications for a state's space power.

The strategic applications of space technology by state actors

As progress in research and the number of actors in space increase, policymakers find themselves confronted with a confusing variety of space technologies that are nevertheless of strategic importance to their national interests, security and social welfare. For the purpose of this book, key space technologies have been classified into four main categories: (1) communications and remote-sensing satellites; (2) space launch rockets; (3) global positioning and navigation systems and (4) new technologies. This section briefly outlines the four technologies and the benefits that each could offer states and their populations. Selected case studies are provided on (a) the use of satellite-based and remote-sensing technology for health and social purposes and environmental monitoring by the Indian Government; (b) how increased launch capacity could affect various areas of Iranian space power and (c) the strategic use of the Global Navigation Satellite System (GNSS) by the European Union (EU) and the geopolitical implications of Galileo.

Communications and remote-sensing satellites

Satellites are acknowledged today as one of the most important technologies in space. In its simplest form, a man-made satellite can be defined as any object made by people and launched into space to orbit the Earth or another celestial body.[39] Satellites can serve any number of purposes, from communications to remote sensing and Earth observation. In addition to the physical satellite launched into space, satellite systems include ground-support stations, which help to launch the satellite, control its movements and oversee communication between the satellite and the ground team.[40] Traditionally, one of the most significant uses of communications satellites and remote-sensing technology was for military purposes, which refers to the sixth capacity of *meta*-geopolitics. The satellites provide military planners with accurate and timely information about the situation and movements of the targets, the topography and weather conditions, and help prevent surprise attacks. They also ensure safe and reliable communications with the troops. Satellites remain a critical means of exploiting the vulnerabilities of the enemy and thus enhancing military power.

The strategic applications of communications and remote-sensing satellites are by no means limited to military affairs. They concern many aspects and have the potential to affect different statecraft capacities. For example, as noted above, they can have significant impact on health care and social issues, the first capacity of *meta*-geopolitics. The use of remote-sensing technology can help predict and monitor the spread of infectious diseases and assist relief operations after natural disasters. Communications satellites can improve health care for people in poor and remote areas by providing access to telemedicine or by enabling people to contact doctors in other locations. By increasing a state's ability to cope with social difficulties, satellites contribute to improving health conditions which in turn reinforces human capital and fosters economic growth and productivity in a country. Improvements in this field, realized with the help of satellite technology, can contribute to strengthening a state's geo-economic standing in the world.

The use of remote-sensing technology is also becoming increasingly valuable for scientists and policymakers dealing with energy scarcity, climate change, environmental degradation and pollution issues – the third capacity of *meta*-geopolitics.[41] Environmental disasters, such as hurricanes, earthquakes or flooding, have huge destructive potential and usually lead to large forced movements of populations, placing massive burdens on local infrastructures and economies that in some cases can result in internal political instability. Remote-sensing satellites can provide early warning, assist urban and rural planning and show changes in coastlines, forest densities and ice caps. Satellite technology thus helps statesmen to gauge potential threats early and take the necessary measures to mitigate or minimize the security or economic consequences of natural disasters.

Finally, satellite technology makes a significant contribution to international diplomacy efforts, the seventh capacity of *meta*-geopolitics. Remote-sensing technology is widely used to monitor arms control and environmental treaties. Despite the risk of extensive foreign knowledge of national territory and national facilities, satellite images can serve as a confidence-building measure and enhance trust and cooperation in the international community. Although satellites are not an absolute or complete method of verification and monitoring, they are becoming increasingly important and being used more frequently.

The purpose of this section is not to provide an exhaustive list of all the potential applications of communications and remote-sensing technologies, which are numerous and diverse. Its focus is the geopolitical implications of satellites that states can use to achieve national goals, be they military, health- or security-related or environmental. Satellites are a highly strategic technology, access to which 'mirrors, reinforces and occasionally transforms terrestrial power-geometries of states in the world system. Although satellites circulate in outer space, their origins and impacts occur very much on the ground'.[42] India provides a concrete example of how communications satellites combined with remote-sensing technology can be used to improve the environmental, social and health capacities of a state.

Case study: Environmental, social and health benefits – a development-oriented space programme in India

Most developing states face the challenge of bringing education, health and information to their most vulnerable, spread over vast and remote areas, which often represent a large proportion of the country's population.[43] Due to its 'coverage of large areas from its vantage point, its ability to bridge the divides irrespective of remoteness of locations and distances, its lack of demands of a prerequisite state of development, its capacity to converge with a range of complementary technologies, its resilience to disasters on earth's surface, and its reliable and rapid response',[44] space technology can help statesmen better address national development needs and empower the most vulnerable populations.

This has been demonstrated in India, which has a space programme that 'has distinguished itself as one of the most cost-effective and development-oriented space programmes in the world'.[45] India realized from the outset that space technology provides countless social and developmental opportunities in addition to its obvious military benefits. Therefore, the Indian space programme had from its beginning a strong strategic focus on the socio-economic development of the country, coupled with a quest for 'self-reliance'.[46] Vikram Sarabhai, who is considered the father of the Indian space programme, said that its guiding vision was to 'be second to none in the development of Space technology and its applications to solve the real problems of man and society'.[47] Education through satellite communication networks, agricultural monitoring, pest early warning systems and telemedicine establishments are just a few examples of the possible applications that communications satellite and remote-sensing technology can offer for societal development. Using concrete examples, this section demonstrates how a focus on the civilian applications of space technology has benefited the Indian population and strengthened the state's power in many areas.

India's space programme started in the 1960s when it began to see the huge potential for space technology as an area for its national development.[48] The Indian space programme, which is run by the Indian Space Research Organization (ISRO), focuses on two main activities: developing space technology for various nationally oriented tasks, such as communication satellites, meteorological applications and remote-sensing satellites for the management of natural resources;[49] and rockets and launch capabilities in order to maintain its independence in space.[50] The programme has been multidimensional from the beginning in the sense that space technologies have been and are used for the various areas that matter to India's development.[51] However, the Indian space programme is best known for the major impact it has had on national social and health matters.

Satellite technology has been used to provide distance education in the most difficult and inaccessible districts of India. For example, the Satellite Television Experiment (SITE) carried out by ISRO in the 1970s connected rural and remote areas to national telecommunications networks and television programmes. The project reached about 2500 villages and broadcast programmes intended for a wide audience on agriculture, health care, hygiene and nutrition issues.[52] Based on the success of this first experience, other projects followed, such as the Training and Development Communication Channel (TDCC) and the launch of Edusat in 2004.[53] The use of satellite technology created educational networks, trained teachers and brought core information about agriculture and health to marginalized populations in remote areas of India.[54]

India was also a forerunner in the practice of telemedicine. India's large rural population and the wide disparities in access to health care made communication satellite technology a priority in the national development strategy. It allows rural clinics to contact tertiary hospitals and receive assistance with diagnostics and treatment through satellite consultations. As a result, in 2009, the INSAT-based[55] telemedicine network connected 235 hospitals –195 district, remote and rural hospitals and 40 specialist hospitals in major cities.[56] The telemedicine industry facilitates medical care in rural areas, and the widespread use of such technologies is making it an increasingly lucrative business for doctors and commercial enterprises alike.

In addition to the health and social benefits, India also uses satellites to deal with environmental issues, to improve agricultural practices and natural disaster mitigation.[57] Whether for the acquisition of data on soil and water resources, the detection of diseases, information-gathering on crops or early warning concerning pests, remote-sensing technology provides many opportunities for agriculture. In addition, weather forecasting satellites combined with India's Cyclone Warning Dissemination System permit better monitoring and prevention of natural disasters. According to Sheehan, '[w]hereas during the 1970s coastal cyclones caused thousands of deaths in India, few deaths were reported once the satellite-based meteorological and early-warning systems became operational'.[58]

India's space programme has also created indirect spin-offs in other sectors. For example, Indian industries have been allocated a considerable role in the implementation of the programme, which has stimulated a national space industry and generated additional income for the national economy. Furthermore, the programme has been developed in collaboration with a wide range of national actors, including academics who, in turn, have produced significant advances in technology and increased the quality of human resources in many disciplines. Finally, the ISRO has established collaborative relationships with other space agencies and international organizations, 'ranging from sharing of scientific data to extending capacity building assistance for space applications and conduct of joint space missions',[59] that have strengthened India's ability to exercise its softpower tools in international forums.

According to Sridhara Murthi, the Indian space programme must be understood from the perspective of the developmental needs of India, a country with 'a large population and mind-boggling diversities in economic, social and cultural aspects and natural endowments'.[60] The use of communications satellite and remote-sensing technology in that particular context with a strategic focus on the civilian applications has contributed decisively to reinforcing state capacities in critical areas. Education is a core element of sustainable development that contributes to the fight against poverty and environmental degradation while improving the health and social cohesion of a country's population.[61] The health of the population is another factor that seriously affects a state's power situation and directly impacts its human capital and national economy. Moreover, the interconnections between the seven capacities of *meta*-geopolitics mean that the Indian space programme indirectly contributes to encouraging scientific research and human development, strengthens an important sector of the national economy and enhances the reputation of Indian technical knowledge on the world stage. In March 2011, India's Prime Minister, Manmohan Singh, reaffirmed his support for India's space programme and clearly positioned tele-education and telemedicine as strategic priorities for the coming years. He also expressed his belief that making space services accessible to all sections of Indian society would help advance sustainable development and reduce disparities, creating a more equal and peaceful India.[62]

Space launch technology

Satellites are useful to our everyday lives, but they would not be able to perform their functions without essential infrastructure which provides crucial services such as launching, fuelling and repairing. The set of technologies that make such core functions possible is known as space transportation technologies, and it includes everything from launch pads to the propulsion equipment and orbit transfer vehicles required for traditional space lift missions, to newer developments in in-orbit refuelling and the servicing and maintenance of satellites and spacecraft currently in orbit.[63] Space launchers were developed from ICBMs, and their main function is to place objects in orbit.[64] The technology used in space launchers and space transportation is therefore interlinked with the technology of satellites and other objects that need to be transported into space. Space rocket launchers are categorized as being used for peaceful purposes but the ability to launch ballistic missiles is a highly controversial issue that has important strategic implications for geopolitics.

Ballistic missiles and space launchers have traditionally been used to reinforce state military and national security capacities (the sixth capability of *meta*-geopolitics). Launchers enable the military to fire long-range missiles and are therefore significant for regional as well as global security. The ability to launch a ballistic missile is often linked to the nuclear capabilities of a state and is central to a state's military power, and therefore a crucial area of interest when discussing the strategic and geopolitical consequences of space launch technology. Military strength has an obvious impact on a state's negotiating capacities at the international level, thereby affecting the seventh capacity of *meta*-geopolitics.

The location of a space launch platform is important because the rotation of the Earth can provide an additional push when launching an object. This additional speed is greatest at the equator when launching towards the east. When launching west, a rocket must compensate for the headwind and add additional fuel in order to reach orbit. The ability to launch as close as possible to the equator is therefore an important cost issue. These aspects are highly relevant in terms of geopolitics. Another important requirement for the location of a launch site is that it be surrounded by a large unpopulated area, such as the open ocean or large uninhabited landmasses. Rockets eject their lower stages and sometimes self-destruct in flight, and such areas are needed in order to reduce the risk in case of an accident. Examples of optimal locations are the Brazilian coast, the coasts of Kenya and Somalia and any of the Pacific islands around Papua New Guinea. For example, although Brazil is not an official launch state, since it does not have indigenous spacelaunch technology, its launch site at Lançamento de Alcântara in Maranhão, located just two degrees south of the equator, has been used to test Brazilian rockets and for Russian space launches.[65] The space base at Alcântara is the spacelaunch base closest to the equator of all the launch centres in the world, and rapid orbit entry and fuel savings for objects being placed in geostationary orbit are powerful negotiating tools for the Brazilian Government in discussions related to scientific and technological cooperation with more advanced space powers.[66]

The satellite launch industry is also of great importance to a country's economic development. Although the revenues generated by the launch sector are smaller than the ones from the satellite services and manufacturing,[67] launchers are of course indispensable to the entire industry. Companies or governments with satellite launch capacities may be able to secure lucrative contracts with customers that want to have objects placed in space, for example, the USD 500 million agreement between SpaceX (the first private company to successfully launch a spacecraft) and Iridium Communications, contracting SpaceX to launch the next generation of Iridium's communications satellites.[68] The commercialization of the space launch industry is progressing, and it is hoped that the private sector will provide launch capacities at a lower cost than governments. In any case, a space launch system can increase a country's national pride and international prestige, leading to positive and unifying effects in domestic politics, the second capacity of *meta*-geopolitics. Indeed, when China first launched two men into space in 2005, Chinese self-esteem experienced a massive boost. Prime Minister Wen Jiabao spoke of a 'glorious and sacred mission' and a 'great achievement' that would go down 'in the annals of the history of the Chinese people'.[69]

The example of Iran below demonstrates how increased launch capabilities affected the various dimensions of its space power, including military and security capacities, domestic politics and national prestige, research and human capital and finally its diplomatic influence. Once again, this is not an exhaustive list of all the possible strategic applications of this space technology. Nonetheless, this case study provides a better understanding of how space launch technology can affect various components of a state's space power.

Case study: The development of launch capacity in Iran

Iran's plans to use 'space and its technologies for peaceful purposes and the nation's welfare' go back to 1958, when Iran joined 17 other countries to found the United Nations ad-hoc Committee for International Cooperation on Space, the future Committee on the Peaceful Uses of Outer Space (COPUOS).[70] The first serious steps towards the realization of a national space programme were undertaken in the reign of Muhammad Reza Shah Pahlavi. In 1977, the Shah supported the creation of an Iran communications satellite system, Zohreh.[71] The 1979 Islamic revolution and the war with Iraq put Iranian space efforts on hold, and Iran had to wait until the establishment of the Iranian Space Agency (ISA) in 2004 to fully institutionalize its space activities and finally see advances in the field of space technology.[72] Today, the ISA has a wide mandate that includes policymaking, manufacturing, the launch and use of national satellites, approval of space-related programmes, identifying guidelines for space cooperation and clarifying the Iranian position in space-related international bodies.[73] According to Parviz Tarikhi, Head of the Microwave Remote Sensing Department at the Mahdasht Satellite Receiving Station, '[p]romoting the applications of space science and technology for peaceful purposes is both a vital part of Iran's current plan and an essential part of its strategy. This includes paying close attention to important concepts such as public awareness, capacity building, research and exchange of experience simultaneously with the expansion of bilateral and multilateral cooperation at the regional and global levels.'[74] Thus, the development of national space technology, capacity building of space experts, as well as increased national pride and international influence appear to be critical strategic goals of the Iranian space programme.

The world closely monitors Iranian efforts to become a nuclear power, but its developments in satellite launch capacities receive far less attention. Missile and launch technologies, however, are complementary to Iran's nuclear programme and should be considered just as important.[75] As Uzi Rubin, one of the pre-eminent analysts of missile systems in the Middle East, argues, '[n]o other country in the world, including established industrialized powers, comes close to Iran in the number and variety of ballistic missiles in development or already deployed'.[76] Furthermore, he points out that Iran's missile and space activities cover almost all the technologies available so far and that the range of the missiles has been steadily increasing.[77] Interestingly, Iran does not make a secret of its launch capacities – on the contrary, its transparency exceeds that of many other powers, and missile launchers are publicly displayed and paraded.[78] In 2008, Iran launched a rocket capable of transporting satellites into space (Kavosh I), followed by its first domestically produced satellite (Omid, Hope) in 2009, effectively making Iran a member of the exclusive spacelaunch club.[79] In terms of the geopolitical implications of Iran's satellite launch capacities, as is noted above, the strategic use of space technology can influence a number of a country's power capacities. In the case of Iran, it has clearly strengthened its posture in the domains of science and human potential, military and security capacities, domestic politics and international diplomacy.

Any space programme relies on knowledge and human potential, research and development and scientific capacity building. In fact, the quality of human resources in a high-skill sector such as space technology is an important measure of a state's space power and its capacity to project that power abroad. Such a highly qualified and reliable pool of space experts is also a prerequisite for national economic growth and international competitiveness. The realization of sophisticated space launch vehicles (SLVs) requires a high level of expertise in systems engineering, conceptual design, integration and testing and simulation and quality assurance.[80] The first Iranian domestic SLV (Safir-2) was developed by the Ministry of Defence and Armed Forces Logistics (MODAFL) in close collaboration with research and academic institutions and foreign countries, including North Korea, Russia and China.[81] This has contributed to significantly advancing Iranian space technology development, building a pool of highly skilled space experts and benefiting other sectors of industry, such as propellant production, navigation systems, cabling and testing.[82]

The state of technological development in a country is also important from a military perspective and Iran's space programme is no different from that of other nations. Despite the fact that Iran's launch technology is being hailed as a peaceful accomplishment, its military usefulness, like all dual-use space technologies, cannot be underestimated. Many governments observe with concern that satellite launchers can be used for non-peaceful purposes, as is noted above, and the range of Iranian ballistic launch capacities is steadily increasing.[83] Another feature is that despite the importance of civilian endeavours within the Iranian space programme, national security space industries have much larger budgets.[84] The successful launch of Omid in 2009, using a domestically produced launcher, clearly indicates that Iran has gained independent access to space and has become one of the world's few launch-capable countries. As a result of entering the exclusive club of space-faring actors, Iran has significantly strengthened its overall geopolitical position on the international stage.

In addition to benefits in terms of know-how, human potential and military capacity, the launch of its first domestically produced satellite has been presented to Iran's population as a significant national achievement.[85] As Tarikhi explains, '[s]tepping into space using an indigenously developed system has provided Iran with a notable and unprecedented national pride'.[86] Iran's newly acquired space launch capacity helped to boost the morale of its people, feed a sense of national identity and foster national cohesion that, in turn, supports the existing political order. Nevertheless, it is important to remember that domestic stability in a country is determined by numerous factors, such as a long-lasting, transparent and accountable political system, cohesive civil society, evenly distributed economic prosperity, access to health and education and the existence of sustainable and legitimate institutions. The strategic use of space technologies such as space launchers and ballistic missiles undeniably has the potential to affect and thereby reinforce the domestic order. On its own, however, space technology cannot lead to internal peace and sustainable development. A space programme is a very expensive activity and as Tarikhi points out, '[i]t can only be justified in Iran if all aspects of the country's space programme are integrated into the social, economic, educational, technical and political life of the nation. Stepping forward in this important way requires international cooperation and collaboration as the main requisite for the success of such an endeavour'.[87]

Global navigation satellite systems

Through a series of satellites positioned strategically around the world, GNSS enables anyone with a GNSS device to identify their location to within a few metres.[88] GNSS research dates back to the 1960s and the development of the first satellite navigation system in history was related to the US Navy's transit system.[89] This system became operational in the 1970s and was the predecessor of the US GPS, which reached its full operational capacity by 1995.[90] GPS is owned by the US military, however, and like most space technologies, it is a dual-use system. The Russian Federation and the EU both have their own GNSS. Russia's GNSS, GLONASS, was launched in 1982 by the Soviet Union and also fulfils dual-use functions. Today, it is being reconstructed with a new generation of satellites called GLONASS-M.[91] The European

Galileo system is designed for civilian purposes and is expected to be fully operational by 2014.[92] China emerges as another actor in the GNSS area. It launched its first Compass satellite in 2007 and is planning to achieve global coverage by 2020.[93]

GNSS, like other space technologies, has the potential to influence several state space power capacities and to contribute to the accomplishment of a state's strategic goals on Earth in several areas simultaneously. Like other space technologies, the military and security applications of GNSS were originally quite pronounced. It is a serious challenge to operate on unfamiliar territory, sometimes at night, without knowledge of the area. Navigation satellites are one of the most common satellites used by the military for tasks such as tracking, weapons guidance, rescue missions and mapping. At a time when warfare can be carried out at long distances between opposing forces, navigation is becoming a crucial part of the conduct of a successful military operation.

In addition to its military applications, navigation systems are today widely used for civilian purposes. In 1983, the US GPS was made partly available for global civilian use.[94] All restraints were removed in May 2000, improving the accuracy of commercial GPS tenfold.[95] The GPS has now been fully integrated into world civilian and commercial infrastructures. Global navigation systems are used today for civil navigation, air-traffic control and to set a global time standard that synchronizes everything from mobile telephones to scientific experiments.[96] Politically, they represent an instrument of global influence and the present monopoly over the GPS by the United States provides it with important negotiating and diplomatic leverage in international affairs. Space-based positioning and navigation applications are used by the commercial sector and the service sector for telecommunications, mapping, precision farming, public safety, surveying and transportation, thereby affecting simultaneously the economic, health, scientific and human capital dimensions of space power. Furthermore, scientists use GNSS in environmental and atmospheric science, oceanography, geology and geophysics, which directly affect the environmental capacity of *meta*-geopolitics.[97] Overall, global navigation systems continue to grow in popularity and sales continue to rise.

Case study: The development of Galileo

This case study examines the EU and its strategic goals in pursuing the building of an independent global navigation system, Galileo. European dependence on the US GPS and the huge benefits of the navigation services market raised questions about the European position in the field of satellite navigation. The importance of this space technology to EU space power and security provided strong incentives for the creation of an alternative, independent navigation system.

In 1998, the EU began looking into the design of an independent European satellite navigation system, envisaged as an alternative to GLONASS and the GPS. The European Space Agency (ESA) publicly launched Galileo with the aim of offering (a) public services covering the needs of the police, the military and the intelligence agencies; (b) free services providing time and location data; (c) commercial services with up to two-metre accuracy for subscribers and (d) safety services such as air traffic control and search-and-rescue assistance.[98] In contrast to the GPS, which is run by the US DoD, Galileo was placed under the joint civilian control of the European Commission and ESA. Yet, despite being launched as a civilian project, Galileo, like most space-based systems, has a range of possible security applications. In October 2009, the EU launched the European Geostationary Navigation Overlay Service (EGNOS), the first pan-European satellite navigation system, which paves the way for Galileo, the full global satellite navigation system currently under development and planned to be operational in 2014.[99] The European Commission's most recent communication on European space policy prioritized the social and economic strategic applications of Galileo. Further, it highlighted the importance of the success of Galileo to achieving EU independence in space, increasing its competitiveness in the global space services market and enhancing EU security and international cooperation in space.[100]

Like other space technologies, navigation systems applications are critical to military as well as national and regional security, and therefore they are considered of strategic importance to the EU. The current US monopoly over the GPS ensures US control over the information gathered and compels other states to conduct vital operations such as air-traffic control, border control, defence and traffic management under the supervision of US military authorities. From an EU perspective, such a situation 'leads to critical dependencies as GNSS are very pervasive, providing vital position, navigation, and timing information for a whole range of daily-life activities and for Europe's security and social and economic development. By relying only on GPS-based applications, the EU would be exposed to the potential non-availability of the GPS signal, which is beyond the EU's control since its primary objective is to support the military operations of a third country.'[101] An independent navigation system will provide the EU with control over its operations and thus protect it against the potential risk of disturbances of the US GPS, which is particularly important in times of conflict. In sum, Galileo can offer the EU independent access to space, thereby making it better able to cope with global security issues. EU independence in space would also improve its position on the international stage and provide competition to US, Russian and Chinese alternatives.[102]

In addition to its security advantages, Galileo has important economic benefits. Possession of its own GNSS will enable the EU to fully benefit from the boom in the navigation services market and substantially reinforce the economic dimension of EU space power. Modern society consumes a wide range of GNSS applications and their number is expected to increase. Given the many GNSS applications, Galileo can offer the EU a unique opportunity to play a key role in the global navigation market, thereby generating new employment opportunities, gains in competitiveness and important economic revenue. According to the ESA, by 2008 sales of GNSS assets and services worldwide amounted to EUR 124 billion, but the EU only had 20 per cent of the market.[103] The development of Galileo 'would make a decisive contribution to the development of a knowledge-based society and the creation of high-value jobs in the EU'.[104] As such, it will give the EU a valuable economic impetus and strengthen its geo-economic standing on the world stage.

With regard to the development of science and human potential, another EU strategic goal, the development of Galileo could lead to significant developments. Satellite navigation is considered a high-technology industry and investment in and development of an independent satellite navigation system would obviously lead to the creation of a pool of highly skilled experts and 'constitute an endorsement of Europe's high technology ambitions'.[105] The development of an independent satellite navigation system will encourage innovation and further advances in that field.

Finally, the development of a programme such as Galileo can also provide a stepping stone in terms of institutional arrangements and therefore affect the seventh capacity of *meta*-geopolitics, which deals with a state's capacity for international diplomacy. The realization of Galileo will require a certain level of interoperability with GPS and 'the European Galileo system should be made interoperable with GPS in a cost-effective way to increase performance for the benefit of service providers, large user communities and individual end-users around the world. The European programme should remain open to international cooperation and partners (such as Brazil, Japan, Turkey and Russia).'[106] Hence, the development of Galileo could provide an incentive for more formal and lasting exchanges and collaboration between the EU and other actors in the global satellite navigation sector.

In the light of the examples discussed above, the development of an independent European global navigation satellite system, encompassing Galileo and EGNOS, offers the EU numerous strategic advantages in terms of security capabilities, economic benefits, technological advances and diplomatic leverage. Galileo can contribute to strengthening the EU's overall geopolitical posture and allow it to challenge US and Russian systems as well as emerging systems in China and India.

Future space technologies

Space technology is a fast-evolving domain and geopolitics is not a static science. The association of these two dynamics leads us to consider a number of new technological developments in space that might affect the power of states and future power relationships between them. By applying the *meta*-geopolitics approach in the sections above, I have provided an overview of the possible strategic applications of three important space technologies and of the way in which they could improve the geopolitical position of two state actors and one community of states. *Meta*-geopolitics is useful for analysing current situations, but its multidimensional analysis also helps make accurate predictions about future trends and the way they might affect geopolitics in the coming decades. This section identifies potential future space developments and how they might influence geopolitics. Rising concern about climate change and energy scarcity means that a large share of space research and development will probably be conducted in the environmental field, the fourth capacity of *meta*-geopolitics. Examples include technological advances leading to the realization of *asteroid mining, space-based solar power* (SBSP) and *non-rocket space launches* (NRSLs).

The depletion of raw materials on Earth is becoming a problem that is seriously affecting the power of states and poses important threats to modern societies. Space exploration has led experts to believe that outer space might offer potential solutions. Satellite technology has increased our knowledge of asteroids. Detailed imaging and lasers show extraordinary variations in their structure and surface. Studies have revealed the presence of a wide range of materials in asteroids, such as iron, platinum, water and trapped or frozen gases.[107] Governments, commercial companies and institutions are now considering ways to exploit these opportunities. Although asteroid mining is so far only a hypothetical idea, it could have important economic potential and has been seriously considered for environmental and developmental reasons.

Another widely discussed technology that could eventually help statesmen cope with the scarcity of energy in the future is SBSP. The Sun shines day and night in outer space, and is unaffected by weather, the seasons and the Earth's atmosphere. According to recent research, '[a] satellite roughly 36,000 km (about 22,000 miles) above the earth can collect the sun's energy using a large solar array and then send a beam to a "rectenna", or ground-based antenna. The energy would then be converted into electricity.'[108] Solar power in space would be much more efficient than solar power on Earth. Solar power from space would provide clean and endlessly renewable energy, and could be crucial in combating climate change and dealing with ever-increasing demands for energy. If the costs can be reduced, SBSP could have important geopolitical implications.

First, it could reduce the political tensions that arise from energy scarcity. If the consumption of natural resources continues at its present rate, the competition for natural resources will increase, enhancing the risk of conflicts over oil, gas and rare minerals. SBSP has the potential to provide the energy needed to satisfy not only an increasing world population but also an increase in demand per capita. SBSP would be particularly important for countries on the Equator, where launch sites can create incentives and opportunities for stability and growth.

Second, it is a clean energy that could help combat the effects of climate change caused by the release of carbon dioxide into the atmosphere, largely from the burning of fossil fuels. It could also reduce dependence on non-renewable energy resources and decrease the incentives for developing nuclear power. An expansion of nuclear power generation has been proposed as a way to decrease carbon dioxide in the atmosphere, despite the risks related to the proliferation of nuclear weapons or the potentially negative environmental effects in the long term. SBSP could replace all these controversial energy sources and avoid potential conflicts and sources of tension in the future.

Energy and environmental security are global challenges, and resource and energy scarcity can pose serious threats to national and global security. Dependence on natural resources is an influential factor in geopolitics and although technical challenges remain, asteroid mining and SBSP could lead to strategic opportunities for enhanced security. SBSP could have significant implications for countries with high energy demands, as well as for countries that produce and sell energy from traditional, non-renewable sources. Research and development in the space industry could result in both new and more efficient sources of energy, and in the development of less energy-intensive devices.

Nevertheless, it will be impossible to mine asteroids or harness SBSP without cheap and reliable access to space.[109] The possibility of using NRSLs is being developed in order to provide a way to reach outer space without using conventional rockets, the only method in use today. As activity in space increases, so does the number of launches. There have been reports of environmental damage around launch sites, such as the Baikonur site in Kazakhstan, where alarm has been raised over the health impact of toxic rocket fuel, and around the Kennedy Space Center, where there is concern over the reduction in the number of plant species.[110] While these problems have so far been limited, an increased market for space launches might have more serious consequences. According to Alexander Bolonkin, an expert in technical sciences, 'in the current political climate new, cheaper and more "fuel-efficient" methods are being investigated. Such new methods include the gas tube method, cable accelerators, tether launch systems, space elevators, solar and magnetic sails, circle launcher space keepers and more.'[111] If the necessary technology could be developed, our space presence could increase significantly at a much more reasonable cost. This would not only facilitate space launches, but also have spillover effects for other potential developments in space, including the asteroid mining and SBSP projects.

New technology is constantly being developed in other areas as well, and new research in material science, nanotechnology and general engineering developments could have a great impact on the way we act in space. Nanotechnology in particular is beginning to have a major impact on space technology, since it enables engineers to make space instruments smaller than before. Developments in materials science can also have spillover effects on space technologies, enabling the development of new types of spacecraft. These emerging technologies are not developed specifically for space use, but they could change the way we use space technologies. Especially interesting is the recent development of smaller satellites. New technology is reducing the size of satellites, which enables them to orbit in constellation groups. This will allow them to make multiple remote measurements in space or in-orbit inspections of larger satellites.[112] These types of satellites will be able to perform new tasks and will lead to reduced production costs by their design and through large-scale production. They will also require smaller and cheaper launch rockets. They could be launched together with other spacecraft and the launch costs could therefore be shared by multiple projects.

Moreover, these new and smaller satellite technologies could create incentives for space activities by non-traditional space actors. It has been argued that small satellites are valuable tools in the development of the space infrastructures and science programmes of developing countries, and that they increase opportunities for these countries to gain access to space.[113] One such example is the South African micro-satellite SUNSAT.[114] The SUNSAT micro-satellite was built by postgraduate students at Stellenbosch University in South Africa and is now in orbit and capturing high-resolution images of the South African landscape for forestry and agricultural purposes.[115] This project has demonstrated that satellite imaging is possible at low cost, and that it can create the basis for further research and development for South African scientists and engineers.[116]

Since the end of the Cold War, international politics and the military focus have shifted, new security issues have emerged and government space agencies are under pressure to reduce their national budgets.[117] Together, nations and private actors are therefore searching for new ways of conducting space-based activities at a lower cost. The development of future space technologies could lead to major changes in the way various actors use space and radically reshuffle future geopolitical configurations on Earth.[118]

Conclusions

Space technology has numerous civil applications. These include improved satellite communications, monitoring of the creeping effects of climate change, the prediction and management of natural disasters, the verification and monitoring of multilateral treaties and the use of global positioning devices to transfer data. Beyond its numerous everyday applications, space technology continues to serve a variety of states' strategically important interests in the military, security, economic, health, domestic politics and diplomatic fields.

The wide range of existing space technologies and their possible strategic applications affects states' power relations and therefore has important geopolitical implications. All space technologies, be they communications or remote-sensing satellites, global navigation systems or space launchers can be employed to strengthen one or more of the seven capacities of *meta*-geopolitics and can therefore help states achieve their national goals. However, a state can be highly influential in certain areas, such as the military and economic, but 'this does not necessarily imply that this actor possesses the complete spectrum of space activities or that this actor can exercise space power to its maximum'.[119] The concrete examples of India, Iran and the EU presented in this chapter enable a better understanding of how improved space power capacities can alter a state's geopolitical posture on Earth. For example, advantages in space technology can enhance national and international prestige and increase a state's diplomatic influence, as is shown in the case of Iran. Space technology can also significantly contribute to developmental goals such as education, access to communications and the provision of general health care in hard-to-reach areas, as is illustrated by India's space programme. Although space technology is an expensive endeavour, it provides significant economic advantages and important technological spin-offs, as is demonstrated in the discussion of Galileo. Moreover, the inherently dual-use nature of space technology means that it also contributes to enhance a state's military capacity and often serves its national security interests.

In addition to the current situation analysis, I outlined future developments in space technology and likely new international trends. The number of actors in space has increased significantly in recent years, and the globalization of knowledge and technology will see more commercial actors and developing countries in space. This will create new opportunities for the new strategic uses of space. The chapter also stressed the potential access to new natural resources. The availability of an infinite amount of minerals and raw materials and the possibility of endless clean and renewable solar power could dramatically change existing relations between world actors.

Science and technology have always influenced foreign policy and international relations, and this is likely to increase in the future. Space technology will continue to affect states' power and world geopolitics as we move forward and increase our presence in space. However, the growing number of actors coupled with continuing research into additional space applications have brought with them new threats. The significant strategic advantages that space brings are only as reliable as the technological assets placed there. Reliance on space-based systems makes the security of objects placed in space critical for all actors. At the same time, as our world becomes more dependent on space assets, the incentives for targeting them increase proportionally, leading to important questions over space security. Chapter 4 addresses the major space-related security issues that international actors are faced with today.

4
Space Security and *Meta*-Geopolitics

In examining the strategic application of space technology by states through the prism of *meta*-geopolitics, I argued that space technology has the potential to strengthen each of the seven capacities of a state's space power and thereby increase its weight in the balance of international power relations. Many actors have acknowledged the benefits that derive from space and decided to take advantage of them, as the United Stated and the Soviet Union did from the start of the space age. Given the heavy financial burden of space programmes coupled with recent budget restrictions, the private sector has emerged as a new key player in the space adventure. As a result, space has become more crowded and competitive while states have become increasingly reliant on technologies placed there for most of their military and civilian activities. These recent developments raise new concerns about security in space. Today's main challenges to space security include a broad range of intentional and unintentional threats, ranging from congestion to debris to potential attacks. Increased reliance on space technology brings with it vulnerability to disruption of space services, while the emergence of new actors in space poses challenges in terms of coordination of orbits, collisions and even potential attack by another state.

It is widely recognized, however, that the preservation of space security is vital to all international actors, and that threats to it deserve a prominent spot on the global agenda. Moreover, there is growing worldwide awareness that the preservation of security in space is crucial not only for national security, but also for global security. In the context of the increased complexity and interdependence of security-related space matters, threats to space security require greater attention from the wider public and from policymakers, in providing coordinated and sustainable solutions to ensure space security.[1]

The extensive use of space opportunities by international actors creates new threats that have the potential to challenge each of the respective space power capacities and threaten international stability. For instance, an intentional attack on an adversary's space assets could trigger chain reactions and produce so much uncontrollable debris that the attacker's own assets would also be damaged and its space capacities reduced. This chapter first examines the nature of these threats, which are classified into two distinct but deeply intertwined categories: intentional or unintentional. It then analyses the effects that such threats could have on the seven capacities of a state's space power and stresses the global and collective nature of the security issues faced by all actors in space. Examples from the seven dimensions of *meta*-geopolitics are provided in order to illustrate these potential risks. For instance, the disruption of satellite services, be it intentional or unintentional, could have potentially catastrophic consequences for military and security aspects (capacity 6) by affecting navigation and communications, on catastrophes management by disrupting environmental monitoring (capacity 4) or on international diplomacy (capacity 7) by interrupting the monitoring of treaty compliance.

The last section focuses on the security capacity of *meta*-geopolitics and argues that security can be affected by space threats far beyond the military and national dimensions. This section investigates the notion of '*national security*' and proposes the '*multi-sum security principle*' as a new approach to conceptualizing the multiple security issues that may result from the irresponsible use of outer space. Finally, this chapter explores potential ways of addressing the challenges to space security and makes a strong case for international cooperation in this area.

Threats to space security

Because of its numerous strategic applications space has become an important arena for all kind of actors, particularly states. As a result, more launches, satellites and actors in space have created an increasingly polluted and overcrowded environment and a heavy dependency on space-based systems. What are the major challenges to space security? Can these space threats affect states' power capacities? Could they put at risk future space exploration and global security? This section attempts to answer these questions. Although a detailed technical discussion of the precise nature of space threats lies beyond the scope of this book, it is important to provide an overview of intentional and unintentional, natural and man-made threats in space, and to examine their possible implications in terms of state power.

Unintentional threats to space security

(1) Space debris

Space is a hostile environment. Access to and a presence in space entail a certain number of risks. One of the serious challenges to a presence in outer space is *space debris*. Man-made space debris can be defined as 'all man-made objects in Earth orbit, including their fragments and parts, that are non-functional with no reasonable expectation of assuming or resuming their intended functions'.[2] Examples of orbital debris include derelict spacecraft, defunct satellites, the upper stages of launch vehicles, carriers for multiple payloads and solid rocket motor effluent.[3] Natural or non-man-made debris mostly consist of fast-moving, rocky meteoroids.[4] Today, the issue of space debris is one of the key security concerns in outer space. Debris can collide with and damage or destroy spacecraft and satellites orbiting the Earth. Moreover, the distribution of debris is not regular in space. The threat to a satellite depends on its orbit, and the regions crowded with satellites are also the most heavily populated with debris and, hence, the most vulnerable.[5] Most space debris is quite small in size, but it travels extremely fast, making any collision potentially dramatic. There are two main causes of orbital debris: the unintentional, such as the break-up of various objects placed in orbit resulting from routine space activity; and the intentional as a result of the testing of anti-satellite (ASAT) weapons. The latter is further discussed in the section on intentional threats. The international community has started exploring possible ways to control the production of debris and clean up space, but no effective regulation or means of removing space debris has been found so far.[6]

Space debris is a growing concern as it can stay at high altitudes for decades. If the number of debris at certain altitudes becomes too great, it will be impossible to use these regions for further space exploration, as debris presents a serious threat to the lives of astronauts and collision with a satellite could cause satellite service disruption and create even more debris. In the long term, debris would then clash with other pieces of debris, thereby multiplying the number of fragments circulating Earth and bringing them to a critical level, 'meaning that collisions between objects are frequent enough that they produce additional debris faster than atmospheric drag removes debris from the region. The additional particles further increase the collision probability in the region, which leads to a slow-motion chain reaction or cascade as the large objects in orbit are ground into small fragments'.[7] This scenario is referred to as the Kessler syndrome, named after a US scientist who first expressed such concerns in the 1970s.

Besides the risks of collision in space, there is also a danger that pieces of space debris may re-enter the atmosphere and hit houses, infrastructure and people. While most space debris burns up on re-entry, some debris is big enough to reach the ground. The chances of space debris causing damage on Earth are low, but there have been cases in the past in which a person has been hit. Re-entering debris may also be composed of hazardous materials. So far, there have been two cases in which radioactive material from radioactive power sources is known to have reached the ground. Such occurrences, should they happen more frequently, may cause fear among populations, and eventually have negative impacts on health and the environment, thereby affecting human security.[8]

Growing awareness of the space debris threat has led to the emergence of a number of national and international initiatives to reduce and control the amount of space debris. At the national level, the United States produced its first set of debris mitigation guidelines in the mid-1990s, entitled *NASA Safety Standard 1740*.[9] It was followed by the US Government's formal debris mitigation standards for space operators, issued in 2000, and the Federal Communications Commission requirements produced in 2004–2005.[10] In 2008 NASA published the first edition of its *Handbook for Limiting Orbital Debris*.[11] In the same vein, the European Space Agency (ESA) published in 1999 a *Space Debris Mitigation Handbook*, revised in 2002.[12] The same year, ESA issued the European *Space Debris Safety and Mitigation Standard*, followed by new mitigation guidelines in 2003.[13] The EU *Code of Conduct for Outer Space Activities* also calls on states to refrain from debris-generating activities and invites them to reaffirm their commitment to the United Nations Committee on the Peaceful Uses of Outer Space (COPUOS) guidelines, discussed below.[14] Japan, France and Russia have also developed their own sets of guidelines. Overall, the national debris mitigation guidelines prescribe similar recommendations, such as the minimization of debris production during 'normal' space activities; the reduction of 'residual propellants, batteries, flywheels, pressure vessels, and other instruments [...] at the end of their operational lifetime. All major national debris mitigation guidelines address the disposal of GEO satellites, typically in graveyard orbits some 235 km above GEO, and most seek the removal of dead spacecraft from LEO within 25 years'.[15]

At the global level, the discussions on space debris started in 1994, within the Scientific and Technical Subcommittee of the COPUOS. In 2001, the Inter-Agency Space Debris Coordination Committee (IADC) was requested by COPUOS to develop a set of international debris mitigation guidelines.[16] These guidelines, the *IADC Space Debris Mitigation Guidelines*, were adopted by consensus in 2002.[17] The guidelines were presented to COPUOS in 2003 and functioned as the basis for that body's deliberations on the same issue. In 2007, COPUOS adopted its guidelines for member states, implementing the IADC's recommendations at the governmental level, and the United Nations General Assembly endorsed them as a set of voluntary measures with which all states are requested to comply.[18] These guidelines are not incorporated into a treaty and are not legally binding. However, the IADC has enhanced knowledge of debris mitigation resolution and the initiative shows that national authorities are taking the threat of space debris seriously. The IADC, together with COPUOS and others, is continuing to monitor debris-related developments in order to further enhance debris mitigation efforts.

(2) Collision risk

In addition to the space debris issue, growing global activity in space has resulted in an amplified general collision risk. Space is *congested* and this situation led to the first ever accidental collision in orbit between two satellites on 10 February 2009. The collision happened between a privately owned US communications satellite, Iridium 33, and the Russian military satellite Kosmos-2251. They collided at a speed of 11.7 km/second above Siberia, destroying both space assets and creating an important amount of debris.[19] This is reported to be the biggest debris-creating event since the Chinese ASAT test in 2007, and the fallout from the collision could potentially threaten the International Space Station (ISS).[20] The US commercial satellite had the potential to make corrections to its orbital position, so there was a possibility of avoiding the collision. It has however been stated that this satellite was not included in the list of US space objects which the US space control system tracks.[21] The US Space Surveillance Network (SSN) detects, tracks, catalogues and identifies man-made objects in orbit around the Earth. This network monitors debris, but it can also keep track of operating assets in space. The SSN can accomplish tasks such as prediction of when a space object will re-enter the Earth's atmosphere, chart anticipated orbital paths and inform NASA of whether an object might interfere with space shuttle or space station orbits.[22] Russia operates the second largest Space Surveillance System (SSS), which uses mostly phased array radars as well as some dedicated radars and optical telescopes.[23] The European Union (EU), Canada, China, France, Germany and Japan are also developing independent space surveillance capabilities.

There are also commercial and academic initiatives on space surveillance. The Space Data Association (SDA) is a non-profit association of satellite operators created by three leading global satellite communications companies (Inmarsat, Intelsat and SES) in 2009. Through its Space Data Center initiated in 2010, it supports 'controlled, reliable and efficient sharing of data' to improve the safety and efficiency of satellite operations.[24] There are also other initiatives, such as the International Scientific Optical Network (ISON) project. Established in 2004, it has 23 observatories in 10 countries, located at different longitudes and latitudes, which forms a centre for observation, planning and data processing, including maintaining a database of space objects.[25]

Most actors in space overwhelmingly agree that preventing collisions is critical to maintaining uninterrupted service, but there is no coordinated effort to monitor all objects in space. If the United States and Russia had shared orbital information about their satellites, the collision in February 2009 could have been avoided. This emphasizes the need for more cooperative space surveillance, but national security concerns have so far taken priority over the need to share and coordinate data. Combining data from different countries and different actors could considerably enhance the accuracy and volume of important information on the situation in outer space, but it would mean that countries would have to disclose classified and tactical information.[26] While the risk of collision is currently moderate, and no significant disruptions are known to have happened so far, the effects of minor accidents allow an estimation of the real impact of an important collision.

(3) Radio frequency interference

In addition to space debris and collision risks, radio frequency interference is becoming an increasing concern, as the demand for radio frequency spectrum and communications bandwidth grows in parallel with the boom in satellite services and applications.[27] Radio frequency spectrum is a natural resource that allows the transmission of information by the use of electromagnetic waves. It is measured in hertz and divided into sections, commonly called frequency bands. Different users and related satellite services share a wide range of frequency spectrum, ranging from hundreds of megahertz to hundreds of gigahertz (GHz).[28] However, only a small portion of this valuable natural resource is available for the exclusive use of satellite communications, which obviously leads to competition for this small part of the radio spectrum.[29] The competition is particularly severe for the segment of spectrum below 3 GHz. Furthermore, there are more satellites operating in the 7–8 GHz section, normally used by geostationary satellites. Increased interest in 'this advantageous frequency and ever closer orbit slots' raises the risk of accidental signal interference.[30]

The International Telecommunication Union (ITU) is the UN institution in charge of coordinating the world's shared radio spectrum and orbital resources, in order to reduce interference risks. The ITU is composed of 191 member states with voting rights and more than 700 industry groups that actively participate in ITU activities but cannot vote.[31] The radio frequency spectrum and the satellite orbital slots are treated as 'limited natural resources that all states have equal rights to use'.[32] The work of the ITU is guided by a legally binding treaty and regulations agreed by all member states, the *Constitution and Convention of the International Telecommunications Union*[33] and a set of *Radio Regulations*.[34] The *Radio Regulations* are managed through the ITU Radio communication Sector and Bureau, which are mandated 'to allocate spectrum and register frequency assignments, orbital positions and other parameters of satellites'.[35] This legally binding framework calls for 'efficient use' and 'equitable access' to the spectrum and orbits and prohibits potentially 'harmful interference' between the radio signals of different countries or companies.[36] Revisions of and amendments to the existing legal tools and regulations are done during ITU Plenipotentiary Conferences and World Radio communication Conferences, held every four years.[37]

The ability of the ITU to manage the regulation of frequencies and orbital positions in an ever more competitive and militarized space context has been questioned. The ITU's rules and regulations exempt military communications. Nevertheless, member states are requested to observe measures to prevent harmful interference and most governments submit their military satellites to ITU procedures.[38] Furthermore, the ITU does not have a binding dispute settlement mechanism for the resolution of interference problems. Instead, such problems are resolved through bilateral negotiations between the parties concerned. Another weakness is that the organization has no enforcement mechanisms and 'member states are essentially expected to comply in good faith'.[39]

The ITU's responsibility for the placement of satellites into geosynchronous orbit has also become a major issue in the geopolitics of space.[40] Many satellites have been placed in geosynchronous orbit in recent years. There is, however, a technical limit to the number of satellites that can be placed in such orbits. The ITU Constitution states that radio frequencies and any associated orbits 'must be used rationally, efficiently and economically [...] so that countries or groups of countries may have equitable access to those orbits and frequencies, taking into account the special needs of the developing countries and the geographical situation of particular countries'.[41] However, in practice, the ITU is distributing orbital slots on a 'first-come, first-served basis', provided the proposed solution complies with the Table of Frequency Allocations in place and no other state contests it.[42]

Resource allocation has direct consequences for which actors have access to space. In 1976, eight countries on the equator came together to sign a declaration stating that geostationary orbit was a scarce national asset and part of the sovereign territory of each of their countries. Although this declaration has not been formally recognized by the United Nations COPUOS, it is the source of much debate in international space policy.[43] In the mid-1980s, developing nations successfully persuaded the ITU to revise its procedures, reserving at least some slots for every member nation to either use itself or lease to others.[44]

Despite the difficulties, the ITU proved to be quite efficient in coordinating the use of spectrum and orbital slots, partly due to the fact that avoiding interference is in the self-interest of the owners and operators of satellites.[45] However, the increased use of the radio spectrum could lead to competitive demands for radio frequencies, which might lead to more overlapping signals, interference, possible collisions and disputes. The US military is the largest user of space and is examining bandwidth dominance. There are reports of growing concern in the United States that information provided to the ITU in order to identify and resolve frequency and interference disputes could jeopardize the security of systems.[46] This potential compromise of the security of satellites makes it difficult to negotiate any amendment to the ITU regulations to include military satellites and communications technology. However, leaving military operators outside the ITU regime complicates the process and could further increase the risks of interference and unintentional jamming.

(4) Space weather

No less dangerous to space security, and completely independent of human action, are the various effects of space weather. Although it appears empty and endless, space is filled with magnetic and electrical fields, energy and other activities that are not visible to the human eye but have serious effects on sensitive scientific instruments and humans.

> Magnetic storm-driven ionospheric density disturbances interfere with high-frequency (HF), very high-frequency (VHF), and ultra high-frequency (UHF) radio communications and navigation signals from GPS satellites. Exposure of spacecraft to energic particles during SEP [solar energic particles] events and radiation belt enhancements can cause temporary operation anomalies, damage critical electronics, degrade solar arrays and blind optical systems.[47]

The Sun causes solar winds, storms or flares and these environmental changes can put spacecraft and astronauts seriously at risk. Space weather events can disrupt communications, computers, power supplies and navigation systems, and expose spacecraft crews to significant radiation. Solar winds can harm the magnetospheric shield, leaving satellites and other objects more exposed to damaging particles. Atmospheric drags can slow satellites down and make them change their orbits, resulting in the loss of satellites.[48] Hence, like space debris, space weather events have the potential to adversely affect the systems we rely on daily. Major magnetic storms can also affect infrastructure directly on Earth, by damaging electric power grids or contributing to the corrosion of oil and gas pipelines.[49] The space weather threat is becoming a growing concern among space experts:

> [W]ith the growth of the electric power industry, the development of telephone and radio communications, and a growing dependence on space-based communications and navigation systems, the vulnerability of modern society and its technological infrastructure to "space weather" has increased dramatically.[50]

In sum, space debris, radio interference or damage caused by space weather represent important challenges for further space exploration and global security. A lengthy or temporary disruption of satellite transmissions would deprive military forces, environmental specialists, doctors, farmers, humanitarian aid agencies and decision makers of the precious information they collect and transmit and could therefore hamper operations in many sectors. In addition to *unintentional* threats, an important part of space threats are *intentional* threats.

Intentional threats to space security: Weaponization of space

The weaponization of space is considered an extension of the military utilization of space, taking militarization a step further and 'getting into the act of destroying of space assets of other states, either by using ground based or space based weapons'.[51] There is as yet no globally agreed definition of what space weapons are. As is discussed in Chapters 2 and 3, most space-related technology is dual use and therefore many non-military technologies and systems designed for civilian purposes are also capable of performing military functions or can be seen as targets. Moreover, states' growing reliance on space technology can be perceived by potential adversaries as a vulnerability to be exploited. Therefore, attacking rival satellites that are engaged in satellite imagery, communications or precision targeting could be an efficient way of using force in a conflict and a short-term solution to protecting one's own space assets. Earth-based weapons are currently the only existing space weapons category, designated as ASATs.

(1) Anti-satellite weapons

ASAT weapons are not new ideas and they have been on the global military agenda for a long time. During the late 1950s and the early 1960s, there were tests of such weapons on both sides of the Iron Curtain. One of the most important ASAT tests, *Starfish Prime*, was undertaken by the United States in 1962.[52] During this test, the US Air Force exploded a 1.4 megaton nuclear weapon, the fallout from which disabled six foreign and US satellites.[53] The Soviet Union had a specific space defence programme early on. This programme included ASAT capabilities, which were intended to combat the threat of US reconnaissance satellites.[54] The Soviet Union's ASAT system was the Co-Orbital ASAT system, which consisted of a missile armed with conventional explosives. The missile would be launched when the target satellite passed over the launch site, placing the ASAT in orbit close to the target. The ASAT would be guided by on-board radar and explode once it was within 1 km of the target.[55]

By the 1960s, however, space weapons and testing had become such prominent issues that there were increasing calls for stronger international regulation. Thus, in 1963 the *Partial Test Ban Treaty* (PTBT) entered into force, banning the testing of nuclear weapons in the atmosphere, in outer space and in or on the sea. The PTBT was followed by the *Outer Space Treaty* in 1967 and the *Anti-Ballistic Missile* (ABM) *Treaty* in 1972, both of which sought to limit the deployment of weapons of mass destruction in space in some form or another. After the US withdrawal from the ABM treaty, which specifically prohibited the development, testing and deployment of anti-ballistic missile systems on the ground, in the sea, the air and space, the *Outer Space Treaty* remained the sole treaty preventing the weaponization of outer space.[56]

However, as Michael Krepon points out, today '[m]any weapons designed for other purposes can be adapted for use against satellites'.[57] ASAT technology can be defined as 'any object or process with the potential to destroy or make inoperable a satellite'.[58] It is generally considered to fall under the umbrella of space weapons, and it can take a variety of forms. Some ASAT weapons produce 'reversible' effects, such as electronic jamming of satellite communications or laser dazzling of imaging satellites, while others are designed to damage the satellite; this is the case of kinetic-energy weapons, high-power microwave weapons or high-power lasers.[59] ASAT technology takes advantage of the fact that the trajectories of satellites can be predicted, making them inherently vulnerable to attack. Which method of attack is used would depend on the specific nature of the target and the origin of the attack. For example, it is relatively easy to jam satellite ground stations and the receivers of an unprotected commercial communications satellite system, but it would be much harder to successfully employ the same type of attack on military communications satellites, as they are protected with many more layers of encryption. The main ASAT weapons are briefly discussed below.

Conventional kinetic-energy weapons

The technology behind the conventional kinetic-energy weapons is pretty simple and less expensive than space-based solutions: a payload is launched on the trajectory of a satellite which is destroyed by physically colliding with it at high speed. Basically, a state with independent launch capabilities can use its rockets or missiles to launch a kinetic-energy weapon. However, more sophisticated targeting technology is required to accurately locate and reach the target.[60]

The US Army started a Kinetic Energy ASAT programme in the late 1980s and early 1990s. Cancelled in 1993, it benefited from renewed interest and funding in 1996. The United States developed a number of ground-based exoatmospheric kill vehicle (EKV) interceptors, to serve, in the first place, ballistic missile defence purposes.[61] In 2008 the United States modified an anti-missile system to enable a Raytheon

SM-3 missile to destroy a failing satellite, USA-193, before it re-entered the Earth's atmosphere. This event followed China's ASAT test against one of its weather satellites and demonstrated the US capability to reconfigure a missile into an ASAT weapon.[62] The experts tend to agree that for its 2007 test, China used 'a vehicle based on a medium-range, two-stage, solid-fuelled ballistic missile'[63] to launch a satellite-killer against one of its own low Earth orbit weather satellites. It is likely that other countries such as Russia, Israel and India are also exploring ways to develop kinetic-energy ASAT capabilities.

Laser weapons

Lasers are potentially a very important area of ASAT development. They are considered ASAT weapons when they are actively used against space assets. The key property of a laser is the consistent, coherent and highly directional output of its beam.[64] Lasers can place intense amounts of light energy and heat on objects that are thousands of miles away, but the effective use of lasers requires a high degree of control and precision in their positioning.[65] Ground-based lasers (GBLs), air-based lasers (ABLs) and space-based lasers (SBLs) used alone or with a space-based relay mirror architecture (where the laser beam is bounced off relay mirror satellites) can all be used as ASAT weapons.[66] Lasers can blind a satellite if they have enough power, thereby damaging the satellite's sensor. If the beam of a high-powered laser can be held on a satellite for long enough, it has the capacity to jam the satellite or to destroy or physically damage it.[67] The United States, China, Russia, Iraq, North Korea, Cuba and Iran have developed capabilities for this kind of jamming, which, nevertheless, continue to face many technical challenges.[68]

New laser technologies, such as 'adaptive optics' can make lasers more efficient and attractive as ASAT weapons in the future, but the engineering to protect satellites advances at a fast pace. With a constellation of space-based lasers in orbit around the world, one could directly hit any missile launch, from short range to intercontinental, with great accuracy.[69] However, since lasers are of limited use in bad weather, this type of space weapon would need to be complemented by air-borne and surface-to-air missile interceptors in order to provide a full missile shield. Lasers are also commonly used in peaceful space technology, for research and for monitoring dangerous space debris. The peaceful uses of lasers can however be turned into offensive weapons. Once again, the dual-use potential of space technology makes it difficult to clarify rules and draft legislation to prevent harmful attacks on space assets.

Jamming

Today, states are using a wide range of means to disable or destroy adversary's assets in space. For example, the disturbance technique or jamming of a satellite's transmissions is not technically demanding and is widely used.[70] Satellites have low signal strength of transmission, which makes them vulnerable to jamming and other disturbance techniques from land-based transmitters. Globalization and the decreasing cost of space technology will probably spread such capabilities even further. Jamming and disturbance of satellites do not create debris, and could therefore be an alternative for states with extensive assets in space. Although disturbance techniques are less efficient than more destructive types, they are a cheap method of interfering with rival states or companies, and many incidents have been reported. For example, the United States is reported to have destroyed six jamming devices used by Iraq. These jamming devices were intended to interfere with the US Global Positioning System (GPS), effectively damaging the ability of the United States to navigate and guide weapons.[71] The Iranian Government is also reported to have repeatedly jammed commercial satellite signals, as has the Libyan Government.[72]

The ITU has expressed concern about the growing number of intentional jamming incidents reported in the recent years. One of the most recent cases concerns the Iranian jamming of European satellite broadcasts.[73] ITU regulations prohibit deliberate interference but in the absence of workable enforcement mechanisms, the cases often remain unsolved. In the case of Iran the ITU redoubled its calls to stop jamming activities, but Iran has not admitted to the jamming and has simply committed to investigate the matter.[74]

Cyber threats

The increasing number of commercial, communications and military operations connected to various types of satellites also raises concerns about cyber threats to space operations. Satellites' ground systems are a strategic target because they are responsible for tracking, controlling and monitoring satellites. These systems, and the links between them and satellites, are all vulnerable to attack, just like any other computer network.[75] The operator-to-satellite link is incredibly important, and disruption to this connection can be disastrous. Satellite hacking can happen in many ways and can have a variety of results. The attack methods and effects range from stealing bandwidth from other users, misusing satellite capacity for propaganda and signal jamming to breaking security codes, the modification of signals, interception and eavesdropping of communications and traffic analysis. Usually, the destruction of the communication between ground and satellite essentially means that the satellite is lost. No matter how they are executed or what their specific result, cyber attacks on satellites and their related infrastructures can have devastating effects on all levels of a state's power and on the security of society.[76]

James Lewis analyses the core threats of cyber attacks against space assets using China and the United States as examples. The US Government often uses commercial satellites for reconnaissance, surveillance and early warning of missile launches. As Lewis notes, '[s]ince it is clear to most militaries that a good portion of the US advantage in combat comes from satellite data, potential opponents like China are searching for ways to interfere with these services from space and the networks that support them'.[77] Lewis maintains that 'attacks against the ground facilities and the networks that control US space assets'[78] would be especially relevant in this context. Interestingly, the vulnerabilities of government systems to cyber attacks are reinforced by the common overlap between commercial and government satellite operations. Governments often accept commercial space firms' security measures without pushing for the higher level they need for real security.[79]

Potential steps to counteract cyber attacks on satellites include militaries adopting more flexible space architecture and space-based systems using other technologies, such as high-altitude Unmanned Air Vehicles and mini-satellites.[80] New technologies in this vein could provide services comparable to satellite and space technologies, and they would effectively make attacks on satellites useless.[81] In order to better prevent cyber attacks against satellites and their related ground operations units, governments should readjust their approaches in order to bring the satellite industry under the umbrella of broader national security infrastructures.[82] Other steps that can be taken 'include the use of robust hardware on satellites, physical security and logical access controls at ground stations, and encryption of the signals for tracking and controlling the satellite and of the data being sent to and from satellites'.[83]

(2) Space-based weapons

Although space-based weapons have long been recognized as very expensive and less flexible than Earth-based weapons, the Administration of US President George W. Bush expressed an interest in placing weapons in space, with the aim of protecting US military and commercial assets in outer space. Space-based weapons can target locations on Earth and other objects in space. However, a space-based attack on another space object can take hours or even days to prepare.[84] Since space-based objects orbit the Earth at very high speed, each object has only a limited window during which to strike a target on Earth. Calculations have shown, for example, that 17 identical laser weapon satellites would be required to ensure complete coverage of a single location.[85] This would make any such weapons system extremely expensive and complicated. The Administration of US President Barack Obama has reoriented US space priorities, and there are no declared space-based weapons at the moment, but many fear that the placement of a weapons system in space might become a reality.

If such a scenario were to happen, the weaponization of space could foster the proliferation of offensive weapons, sparking a new and potentially deadly arms race.[86] The placement of weapons in space would create tensions and incentives to intentionally harm assets in space in order to weaken the military and civilian capabilities of an adversary. Placing space weapons in Earth's orbit would be a cause for concern to all nations under the flight path or orbit of such weapons and could also cause serious economic damage to them all.[87] US industry in particular would be negatively affected by damage to space infrastructure linked to space warfare because of the important revenue it receives from global space commerce. The areas in which space technologies are used are increasing, and so is the number of countries using such technology. Although the United States is still the dominant player in space, many other countries would suffer from a war in space, making it a very expensive event. Moreover, attacking satellites and other assets in space would create more debris and radiation, thereby increasing the unintentional risk. This close relationship between intentional attacks by hostile states and related unintentional threats such as space debris or the risk of collision is one of the reasons why prevention of the weaponization of space must be a priority issue in arms control discussions. Currently, there are no weapons in space, but the implications of a weaponized space could be devastating and the prevention of an arms race in space is essential to preserving safe access to space and global security.

Space security and the seven dimensions of space power

Chapter 3 discusses the many ways states use space technology to achieve their strategic military and civilian goals. It illustrates how space technology can be used for telemedicine and tele-education, climate change monitoring and observation of agricultural developments, and for communications and banking worldwide. Yet, while space offers numerous opportunities, the present chapter shows that such an extensive exploitation of space technology brings new risks that can adversely affect all seven of the capacities of state power and threaten international stability.

I briefly outlined above the major existing intentional and unintentional threats to space security, and stressed the interrelations between them and the major risks they present to satellite services, spacecraft navigation and, more importantly, to the preservation of peace and stability in space. While the majority of these risks remain theoretical, thus far, some have already materialized, such as satellite collision or ASAT weapon tests. What would be the implications in terms of space power if a specific space technology became unavailable due to natural causes or man-made interventions? What would be the consequences for the geopolitical structure and global security? Would all states be evenly affected by the disruption of their space-based services?

Space-faring and new entrant nations may have different space policies and strategic programmes, but most have important military and economic dimensions. Consequently, disruption to a critical space-based system would certainly negatively affect the military and economic functions of any state. Moreover, the interrelations between different dimensions of space power mean that deterioration in one would trigger a chain reaction, cascading adverse effects to the other components, which might lead to the weakening of a state's overall geopolitical position. Nevertheless, the severity of any damage, if such a scenario were to occur, would vary among states depending on a number of factors, such as the duration of the disturbance, recovery capacity, the technological assets involved, the vulnerability of a state to the disruption or loss of a space-based system and the relative importance of the affected capacity in the specific national context. This section analyses the way in which the intentional and unintentional threats identified above could affect the different capacities of a space power and, as a consequence, increase its vulnerability.

Space debris, collision and frequency interference risks

Our growing presence in space has increased the amount of space debris and the risk of collision and frequency interference, presenting serious threats to spacecraft and in particular to satellites. Today, satellite technology is used by almost every country in the world and supports a wide range of civil and military services, including vital information on positioning, navigation and the time. The all-pervasive nature of satellite technology has created critical dependencies on such systems. As this reliance grows, so does vulnerability to the loss of space assets. Yet, satellites are very fragile objects that can be harmed in many ways, as shown in the previous section. In 2010, the US military's Joint Space Operations Center (JSOC) issued about 1000 warnings that a space asset was at risk of collision. The result was that collision was avoided by changing the position of the satellite.[88] A temporary disruption in satellite services, permanent damage or the total destruction of a satellite could have serious implications for the seven capacities of a state's space power.

Satellite technology has traditionally been used to strengthen the military capacities of a state and has become a critical tool for nearly all military operations and weapons systems. In case of space information degradation or denial, the military capabilities of a space power and its national security could be seriously affected.

Space-based assets are also critical to most daily economic activities. Today, world financial institutions rely heavily on satellite technology for communication services, credit and banking transactions, safe document storage and global trading operations, among other things. Accurate and seamless timing are essential for many digital networks and the availability of Coordinated Universal Time (UTC) is based entirely on satellite navigation systems, in particular the US GPS. A disruption to or loss of the satellite signal would result in 'clock drift', which could affect financial markets, telecommunications, digital broadcasting and the Internet.[89] Furthermore, space-related services have become a major part of the world economy and of commercial sectors. Given the importance of this market, economic losses resulting from satellite disruption could be huge and severely affect the economic situation of space-faring nations.

The disruption or loss of a satellite signal can also affect the environmental component of space power. For example, the disruption of the GPS signal could hamper the ability of the United States to monitor the Earth's upper atmosphere and provide timely alerts of disturbances in the ionosphere. These disturbances can knock out power grids and expose aircraft on polar routes to dangerous radiation. Without GPS, this essential warning capability would be significantly degraded, causing important financial losses and putting human safety at risk.[90] The international diplomacy dimension could also be weakened, as the risk of satellite disruption could reduce the confidence of states in the monitoring and verification systems for climate change and other international agreements. Parties will be more reluctant to engage in multilateral treaties and international diplomacy might be reduced, particularly in highly vulnerable fields. In addition, satellite technology is used for health early warning systems and epidemiological purposes as part of a strategy to improve the efficiency of interventions in natural disasters and post-crisis management.[91] In a health emergency or a post-disaster situation, where real time interventions are of vital importance, even the temporary unavailability of a satellite signal could reduce the efficiency and speed of relief operations and have serious implications in terms of national health and human lives.

Despite the existence of back-up systems, a long-term loss of space assets could have adverse effects on the seven capacities of a state's space power and reduce security on Earth.[92] Therefore, regulating the use and traffic of satellites and other objects in space is in the interests of all space powers. The need for a multilateral surveillance system is obvious, but national security concerns prevent states from extensively sharing orbit positions, speeds and other related information. This remains a key contradiction in the global space debate, and states are still working to strike a balance.

Space weather events

States' critical infrastructures and systems rely heavily on communications, navigation, timing, meteorological and other services provided by space-based assets. Space weather is therefore a serious risk for these systems as it has the potential to interrupt them in space and also directly on Earth.[93] The possible adverse effects of space weather are well-known today and comprise 'power grid outages, high-frequency communication blackouts, interference with Global Positioning System (GPS) navigation signals, spacecraft anomalies'.[94] The negative consequences of these technical failures could be amplified due to the interconnectedness of the structures we rely on and the interdependencies of the seven capacities of *meta*-geopolitics.

Like debris, interference and collision threats, space weather events represent a risk to satellites as they can interfere with satellite signals. For example, satellite navigation systems are extensively used by the military and any long-term disruption of the GPS signal would undoubtedly undermine a state's military capacities. GPS use has grown far beyond its military applications. It is used today for many civil purposes, such as land surveys, weather forecasts, telemedicine, time synchronization and search and rescue operations. It has also been extended to the commercial sector and is widely used by investment banks, mobile telephone companies and computer networks, among others. Consequently, disruption of the GPS would create a chain reaction and simultaneously paralyse many areas of our daily lives, adversely affecting almost all dimensions of space power.

Such a scenario has occurred and caused important damage in many sectors. In January 1994, the outage of two Canadian telecommunication satellites caused by exceptional electron fluxes disrupted communications countrywide. The first satellite was repaired within few hours but the recuperation of the second one took six months and cost USD 50 million to USD 70 million.[95] In July 2002, a solar storm caused the disruption of satellite data transfers, altering the data provided by a weather satellite.[96] In 2003, a severe geomagnetic storm diverted flights to avoid a radiation zone and communications outage, resulting in additional costs of USD 10,000 to USD 100,000.[97] Like the other threats to space security, space weather events can have severe economic consequences.

In addition to satellite disruption, space weather can also threaten space exploration and the human presence in space, thus challenging further expansion by space-faring nations and affecting the scientific and economic dimensions of a state's space power. The United States, for example, has a permanent presence in space on the ISS and regularly exposes its spacecraft and astronauts to space weather threats. In 2003, bursts of energy from the Sun created intense solar particle events and caused geomagnetic storms.[98] This phenomenon resulted in a high-level radiation storm which forced 'NASA officials to issue a flight directive to the ISS astronauts to take precautionary shelter'.[99] Moreover, the same weather storm is believed to have been responsible for the loss of the USD 640 million ADEOS-2 spacecraft, with the USD 150 million NASA Sea Winds instrument on board.[100] Heightened safety risks coupled with a number of expensive losses could result in a slowdown of investment in space research and development.

In addition to the implications for space-based systems, space weather has the potential to affect states' critical infrastructure directly on Earth. The blackout that resulted in outages of electric power equipment through Quebec and the north-east of the United States in March 1989 is commonly quoted as an example of such a scenario.[101] The increasing interconnectedness and complexity of the infrastructures we rely on mean that the impacts of a space weather event on Earth have the potential to affect all and to paralyse states' core capacities. The consequences of such an event could be devastating as the negative effects would cascade through all sectors of society, affecting almost all dimensions of space power. As explained by the National Research Council in a workshop on *Severe Space Weather Events and their Societal and Economic Impacts*,

> [The] [c]ollateral effects of a longer-term outage would likely include, for example, disruption of the transportation, communication, banking, and finance systems, and government services; the breakdown of the distribution of potable water owing to pump failure; and the loss of perishable foods and medications because of lack of refrigeration. The resulting loss of services for a significant period of time in even one region of the country could affect the entire nation and have international impacts as well.[102]

Space weather therefore represents a serious threat to space-based and Earth-based critical infrastructures that needs to be urgently addressed by the international community. This concern has been raised by many space experts, who warn that

> [s]ervice disruptions of relatively short or conceivably very long duration may spread from a directly affected system to many other systems due to dependencies and interdependencies among, for example, electric power supply, transportation and communications, information technology, and government services. As systems become more complex and adaptive over time, the social and economic impacts of space weather are likely to increase.[103]

The weaponization of outer space

The threats to the human presence in space and to space-based assets are not limited to unintentional threats. The dual-use nature of space assets means that most of the technologies, such as satellites, remote sensing technologies and GPSs, can easily become weapons or targets used to weaken the rivals' space power capacities.

For example, because satellites orbit the Earth in fixed, predictable patterns they can make easy targets for attacks. There are many intentional ways to disrupt satellite signals: satellites can be destroyed though ASAT attacks from Earth, signal transmissions can be jammed, satellite sensors can be blinded by a laser or satellite's Earth-based systems can be damaged using cyber attacks. Reconnaissance satellites are particularly vulnerable to intentional threats because of their common military application. They form part of many strategically important military missions, aimed at gathering information on the military and industrial capabilities of other countries, mapping terrain and choosing targets during a conflict, which makes attacks on them highly disruptive and damaging.[104] Although current generations of satellites are designed to defend themselves, their defence systems are not sufficient to guarantee survival against an attack. Consequently, the need for protection could result in the placement of weapons in space. Such a unilateral defensive option was advocated by former US President George W. Bush in his National Missile Defense programme, aimed at building a missile defence shield in outer space capable of shooting down nuclear missiles.[105] If such a system were ever deployed, the fact that states would not be equally vulnerable to a nuclear attack would profoundly alter the balance of power between nuclear powers, which could have disastrous consequences in terms of global security.

It is worth saying here that intentional threats in space are often related to the notion of asymmetric warfare, which can be defined as 'countering an adversary's strength by focusing on its weaknesses'.[106] This means that opponents of a strong space-faring nation could make use of non-conventional weapons and unconventional strategies to exploit the vulnerability that space technological superiority and dependence generate. International awareness of US space capabilities and its resulting vulnerability to these systems' failure has prompted some opponents to use their asymmetric advantage against US space assets.[107] For example, Russia and China cannot compete with the United States in missile defence or in nuclear and conventional weapons. They do, however, have the ability to engage in asymmetric responses through their respective ASAT programmes, which are technically less challenging to construct and use but can enable a weaker opponent to prevail.[108] In January 2007, China made international headlines and caused major diplomatic waves when it initiated a successful ASAT test attack against one of its own low Earth orbit weather satellites, thus becoming the third country to achieve this objective.[109] On the one hand, this test served to show China's military capacities in the international arena. On the other hand, it created enormous amounts of space debris, with one of the 3000 pieces passing close enough to the ISS in April 2011 to raise serious concern about its security.[110] It has also been perceived as a threat by many states and a reminder that the weaponization of space is not only a theoretical risk.[111] The most affected states were those with a high degree of vulnerability to the loss of space-based services, such as the United States, European countries and Russia. Therefore, shortly after the Chinese ASAT test, on 2 February 2008, the United States launched a missile defence interceptor to destroy a US low Earth orbit satellite, thereby demonstrating its space power and military capacity.[112] Some think that future warfare will almost inevitably imply asymmetric means and attacks on or by space-based assets, and this possibility should be prominent on the international agenda.

Because space-based systems work so well today, their beneficiaries take them for granted and pay little attention to them. It is only when space-based services 'are interrupted that we realize the full extent of our dependence on them. [...] we now have commercial remote sensing satellites, direct-to-home satellite television, mobile satellite radio, next-generation weather satellites that are extremely capable, and explosive growth in our GPS markets [...]. Our dependence on space services has compounded by an order of magnitude, bringing significant added benefits but also significant added risks'.[113]

All space assets are equally threatened by unintentional dangers, no matter which country owns them. Providing adequate physical protection for space assets is difficult and sometimes more costly than the asset itself, leaving most objects in space in a state of extreme vulnerability. To address unintentional threats, solutions can be sought in multilateral cooperation, sharing information, the creation of a standardized system and responsible action in space. All space-based services are deeply interwoven with a nation's social and health issues, culture, politics, economics, environment, scientific progress, diplomacy and security. Being of strategic importance to international actors, in particular states, space assets have become a strategic target for potential adversaries. Moreover, increasing national over-reliance on space technology means that states can gain substantial geopolitical advantages by denying, disrupting or destroying rival capabilities to access and use space. Nonetheless, developing space weapons would be expensive and of questionable effectiveness in terms of both offensive and defensive applications. Indeed, an outbreak of space warfare would have devastating consequences for multiple satellite-dependent industries, countries and interest groups around the world.[114]

For example, the impact of the disruption of a GPS signal might be minimal, in case of a quick recovery, but it could seriously reduce the effectiveness and efficiency of critical state operations and affect national security, as well as the environmental and economic dimensions of a state's space power.[115] That said, an intentional disruption of the US GPS navigation satellites for a certain period would not only affect the United States, but have implications for all actors who depend on GPS services. In a globalized and interconnected world, with the economies of the major space powers deeply interlinked, the deployment of new military technologies in outer space would have massive consequences for global security, the international economy and the balance of geopolitical forces. However, it is important to remind ourselves that the geopolitics and security of space are informed by international relations on Earth, and it is on Earth that these issues must be addressed. The potential military use of space would exacerbate geopolitical tensions, cause conflicts and make space impossible to use for present and future generations. The section that follows therefore investigates the relationship between space security and global security and draws some early conclusions on how to address these pressing security issues.

Space security and global security: The *'multi-sum security principle'*

Modern societies are witnessing the emergence of new, global threats such as cyber attacks, energy scarcity, global warming, human trafficking, terrorism and economic turmoil. They have grown in importance with the increasing levels of interconnectedness and interdependence of our globalized society. Moreover, these threats can come from non-state actors and often emerge within the sovereign borders of a state. Non-state actors such as transnational organized criminal networks, corporations, non-governmental organizations and the media are crucial when considering security issues today. Moreover, these security threats not only affect the territorial integrity of the state, but also endanger the state's population and its well-being. In the context of increased complexity and the shifting of security-related patterns, how should the concept of *security* be understood? What is the correlation between security on Earth and security in space? To what extent can the latter influence the global security debate?

The notion of security needs to be placed in a wider context, as part of *'global security'*, where it is no longer just a national concern, but a variation of referent objects as well as the means to address it. In order to create a more inclusive approach to global security, a new principle of security was previously proposed – the *'multi-sum security principle'*.[116] The security challenges that states face in space affect global security within its five dimensions: human, environmental, national, transnational and transcultural components and can have a direct impact on security conditions on Earth. This section briefly outlines the five dimensions of the *'multi-sum security principle'* and highlights their relevance to space security, see Figure 4.1 below.[117]

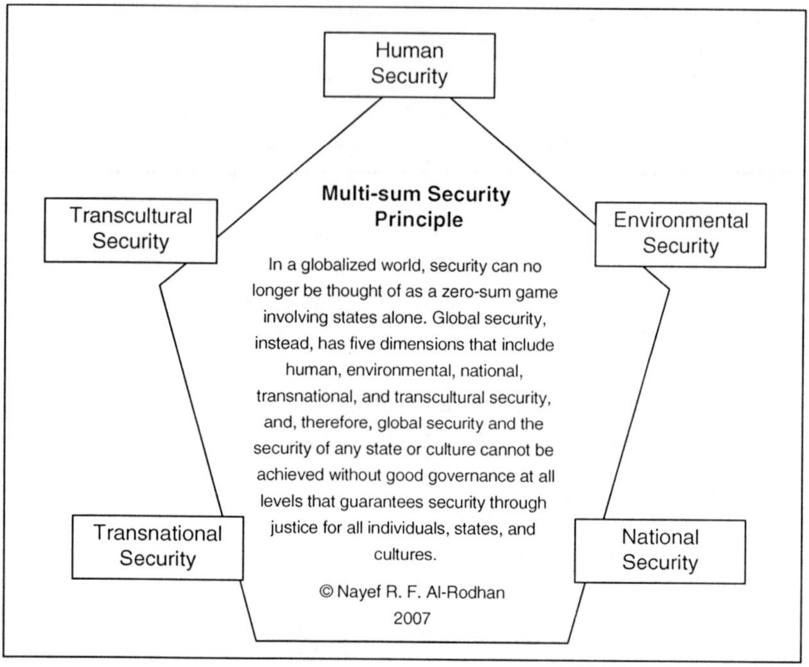

Figure 4.1 The '*multi-sum security principle*'
Source: Nayef R. F. Al-Rodhan (2007), *The Five Dimensions of Global Security: Proposal for a Multi-sum Security Principle* (Berlin: LIT), p. 31. Reproduced with permission from LIT Verlag.

Human security

The first dimension of global security is human security. Placed on the international agenda by the United Nations Development Programme's 1994 Human Development Report, *New Dimensions of Human Security*, the concept of human security calls for a focus on the individual rather than the state, and a people-centred approach to security. The 1994 report defines human security as 'freedom from fear and freedom from want'.[118] One of its key underlying assumptions is that threats to an individual's security can emerge from a state within its own borders and, therefore, a focus on state security is not sufficient. Space technology can have a significant impact on human security at the micro level, if used correctly. On the other hand, it can also pose threats to human security. In the 1990s, the potential impact of space technology on human security at all levels – resource management, disaster management, health, education and so on – received a great deal of attention, eventually resulting in the *UN Vienna Declaration on Space and Human Development*.

This agreement suggests a blueprint for the peaceful use of outer space, with a strong focus on human development and security.[119]

Environmental security

The second dimension of global security refers to the environment, the relevance of which to international stability is increasingly acknowledged – as is its close relationship with other dimensions, such as the national or human elements. Environmental security faces a wide spectrum of threats, such as global climate change, water and air pollution, natural disasters and the degradation of scarce resources. All of these are transnational in nature, as they can affect many countries at once. Remote sensing technology brings significant value in promoting the sustainable use of the environment, monitoring potentially harmful changes and improving risk management and disaster planning. However, its dual-use capacity can transform satellites into perfect targets. Intentional attacks on such assets, or unintentional threats such as debris, the risk of collision or space weather, have the potential to disrupt these critical space-based systems and deprive states of the essential information they provide to ensure environmental security.

National security

National security has traditionally been the primary object of a state's security considerations. It encompasses an important military dimension but the political, economic and societal components are also essential for maintaining security within the sovereign territory of a state.[120] Although no longer the only referent, national security remains a major concern of states. Unlike the Cold War era, when space matters mostly concerned the superpowers, today all countries are turning to space programmes to enhance their military dimension and protect national security. Indeed, many states observed the growing role that space and satellite technologies played in the 1990 Persian Gulf War and have recognized the advantages that such technologies provided to the United States. Influenced by this observation, many states have started to invest in their own space programmes, focusing not only on military technology, but also on other security-related space services such as remote sensing and communications. According to James A. Lewis from the Centre of Strategic and International studies (CSIS), the use of space technology has marked a new direction for future warfare. Overall, Lewis identifies three main reasons why state actors are striving to use space for national security: (1) the political prestige and international influence space activities bring; (2) the technological capabilities that space programmes create or reinforce and (3) the informational advantage space assets and services can bring to military and security operations, and to strategic planning for security.[121] Given the critical role space technology plays in preserving states' national security, intentional or unintentional threats to states' critical systems in space could directly affect their national security dimension.

Transnational security

Increased commercial, economic, financial and personal interdependence and interconnectedness in the twenty-first century mean that a number of new threats to security have emerged that are not national, but transnational in nature: 'transnational security threats refer to security challenges that transcend international borders and threaten the political, social, or economic integrity of a nation or the quality of life or livelihood of its inhabitants'.[122] Today, transnational challenges to security are posed by organized crime, human and drug trafficking, cybercrime or international terrorism. As applied to space, our growing dependence on space-based systems and high vulnerability to their loss can be instrumentalized by transnational actors such as terrorists or cyber hackers who, through asymmetric measures, can seek to damage space operations and use space assets to their advantage.

For example, the spectrum of cyber attacks against space infrastructure is broad, and the impacts can be extensive. Successful cyber attacks against satellites could put at risk intelligence gathering, or undermine key communications structures, provoking devastating results from military and civilian perspectives. In terms of a terrorist attack, according to N.-L. Remuss from the European Space Policy Institute, 'terrorists are both motivated and capable of conducting a terrorist attack in space'.[123] Space security studies, she emphasizes, are lagging behind terrestrial security studies in the sense that the risk of a terrorist attack in space is largely neglected, even though there have been examples of terrorist attacks on space assets in the past. Possible terrorist attacks include attacks on satellites, attacks on launch facilities and ground stations, or attacks on user/service equipment.[124] For instance, in 2007 the Falun Gong movement, which is considered a terrorist group by China, took control of nine central television stations, as well as ten provincial ones, and interfered with the normal broadcasting of programmes by causing disruptions and broadcasting their own propaganda material.[125] The consequences of terrorist attacks suggested by Remuss, such as incapacitating the armed forces, and mass panic, and hence mass casualties and long-lasting psychological effects, seem somewhat disproportionate, but transnational security certainly is an aspect that policymakers should consider when addressing space security issues.

Transcultural security

Closely linked to the aspect of transnational security is the notion of transcultural security, which refers to 'the integrity of diverse cultures and civilizational forms'.[126] In a globalized world, characterized by intensive movements of populations and the coexistence of different systems of belief and political order, cultural issues deserve a prominent spot in any security policy. One of the main threats to transcultural security occurs when diverse communities are not positively integrated into the broader society in which they live. Space technologies, such as satellite broadcasting, can be misused and used for propaganda purposes by either dominant groups or isolated cultural entities to defend their rights.[127] This concern was raised in 1997 by N. Jasentuliyana and K. Karnik, who wrote that 'space communications, for example, can have a potentially negative role in serving as a "weapon of mass indoctrination"'.[128] The example given by the authors discusses the way satellite broadcasting has been used to promote occidental and consumption-oriented lifestyles, culture and values that tend to homogenize societies rather than preserve different cultural identities. Like transnational networks, particular cultural groups might also use space assets and systems to defend and promote their cultural identities. For instance, in 2007, Sri Lanka's Tamil Tigers hijacked a US-based Intelsat satellite, broadcasting its propaganda for an independent Tamil homeland across the Indian Subcontinent.[129] Sensitive issues such as religion and social and moral traditions must therefore be taken into account when addressing space security issues.

In the light of these examples, we can say that all actors in space share complex and intertwined security threats that have the potential to affect each of the five global security dimensions on Earth. Free access and a lack of territorial boundaries in space mean that an attack on an object could easily spill over and reduce security for all in space, including threatening the attacker's own space objects. Moreover, improvements in one of the dimensions can undermine the security in one of the other four areas. For instance, a security assessment applying the '*multi-sum security principle*' reveals that through the weaponization of space, states seek to increase the military capacity of their space power and protect national security. However, such acts are likely to have negative effects in other dimensions of security: environmental and human, through the creation of debris or an increased risk of collision.

It has been acknowledged worldwide that space-related risks need to be mitigated; the question of how to address space security threats remains a difficult and much-debated issue. The section that follows addresses this question. Opinions range from support for peaceful cooperation to advocacy of competition or the unilateral domination of space as a means of containing threats. The issue of cooperation and competition in space and the implications for how to face potential threats are among the major fields of debate between space scholars and practitioners.

Addressing threats to space security

As more actors are present in and dependent on space assets, ensuring that space objects can operate without interference or harm for the benefit of mankind becomes critical to preserving global stability. Probably the most pressing issue in terms of unintentional threats is space debris and the challenge of avoiding future collisions. As our dependence on global space services increases, the number of objects in space will grow, dramatically increasing the probability and frequency of interference or serious collisions with other satellites or debris. There are also many intentional threats, and ASAT weapons figure prominently among them. Threats to space objects have the potential to amplify existing or create new terrestrial tensions. The development of an ASAT capacity by a state can generate suspicion and encourage other actors to develop a similar capacity. The proliferation of such unilateral defensive measures would 'increase the chances of dangerous misinterpretations, especially in the absence of clearly stated policies and meaningful communication between countries'.[130] If no action is taken, we will create more international tensions and might not be able to make use of the precious opportunities that space offers in the future. As Sergei Ordzhonikidze puts it,

> The increase of space debris and the development of space-related weapon technology threaten space security and increase the likelihood of space becoming a more hostile environment. States must appreciate their shared vulnerability in outer space and understand the mutual benefits of guaranteeing peaceful access to space for all the world's peoples. If not appreciated and understood, stability in space, and thus space-based assets, would be seriously threatened.[131]

In the light of this warning, how can risks in space be mitigated? What steps need to be taken and what measures are needed to ensure a peaceful environment in space? In the existing literature, a large part of the security-related debate focuses on the potential weaponization of space and opinions about how to cope with space weapons obviously differ. Chapter 1 briefly flags the origins and arguments of the main schools of thought in space security as identified by Moltz: space nationalism, global institutionalism, technological determinism and social interactionism as well as technonationalism.[132] Each prescribes different recipes to cope with threats to space security, and, depending on the school of thought with which they align, academics and policymakers advocate different approaches. They can roughly be divided into two camps: the pro-weaponization *'space hawks'*, or the *'military realists'* camp, which supports competition and even the unilateral domination of space as a means of containing threats; and the *'space sanctuary'*, *'space doves'* or peaceful collaboration camp, which advocates non-weaponization and the common use of space.[133]

Proponents of the weaponization of space believe that it is inevitable and argue that the United States must do it first in order to maintain its military superiority. Weaponization advocates are also driven by the fear of what they perceive to be the increasing vulnerability of US space assets to a pre-emptive attack, something the 2001 Rumsfeld Commission dubbed a potential *'Space Pearl Harbor'*.[134] At a more pragmatic level, supporters of space weaponization believe that there are important financial gains to be made from dominating this branch of the space industry.[135] In general, this group is sceptical of international commitments on the weaponization of space and believes that the United States should not voluntarily limit its options. As Jeff Kueter, President of the George C. Marshall Institute, puts it, 'Russian and Chinese efforts to advance a new treaty provide cover for their self-serving attempts to constrain the United States, while doing nothing to restrict their own clandestine ASAT programs'.[136] However, US weaponization advocates are not the only ones to think this way and the same ideas are widespread among Chinese proponents as well.

Other policymakers decisively oppose the further militarization of space. For example, Lieutenant Colonel Bruce M. DeBlois proposes a policy of space *'sanctuarization'* and advances many arguments against the weaponization of space. Some of his stronger contentions include the fact that because of the high level of vulnerability of space assets, space weapons would essentially make existing space technology useless, and result in an expensive waste of resources. Furthermore, space weapons are seen as provocative because dominant offensive weapons encourage pre-emption against them, and thus the risk of a space weapons race is high. On a similar note, it is believed that because they are remote and offer plausible deniability, space weapon strategies are inherently escalatory. Since space weapons are typically unmanned, they are easier to use and have less risk attached to them than traditional weapons. Finally, space weaponization strategies are considered expensive. What can be done with space weapons can also be done from the air – without the political baggage of the weaponization of space.[137]

Overall, taking into account the implications of space warfare for global security, it is increasingly important to seek diplomatic solutions to prevent an arms race in outer space. The possibilities: 'range from treaties that would ban anti-satellite and space-based weapons to voluntary measures that would help to build transparency and mutual confidence'.[138] Most international players believe that the weaponization of space would be inappropriate, destabilizing, militarily ineffective and extremely costly.[139] Nevertheless, since the mid-1990s, the establishment of a new multilateral legal regime has encountered serious obstacles. While a large part of the international community opposes the weaponization of space, its persistent national-centred security view of space continues to hamper the emergence of a global binding legal framework.

The main treaty regulating activity in outer space, the Outer Space Treaty (OST), and its supporting agreements are more than 40 years old.[140] While there is a consensus that existing norms and legislation on space need to be revised in order to account for new developments and new space security threats, there are important divergences over how to fill these gaps.[141] Some propose to revisit and reformulate particular treaty articles in a targeted way.[142] Others stress the importance of confidence-and security-building measures (CSBMs) or highlight the role of the private sector and commercial regulations in future agreements.[143] Still others advocate a *rules of the road* approach or the creation of so-called soft law treaties.

Addressing military aspects is only a part of the picture. There is no doubt that the question of the weaponization of space is a major and substantial aspect of the space security debate. However, as Moltz points out, a new understanding of space security is needed, and any tool to address security in space should go beyond military- and weapon-related threats. Space debris, collision risks, frequency interferences and space weather represent global threats to space security that need to be addressed at the international level. Efforts have been made to protect space assets from unintentional threats. Existing instruments include ITU technical guidelines, national regulations, COPUOS guidelines and handbooks developed by space agencies. Since the mid-1990s,

Russia, the United States, the ESA, China and Japan have all developed their own standards aimed at debris mitigation.[144] For example, the US Air Force has tracked space objects since 1957. Both the US Space Surveillance Network and the Russian Space Surveillance System register space objects.[145] While there is a clear need for more cooperative space surveillance, national security concerns have so far taken priority over the need to share and coordinate data. Combining data from different countries and different actors could considerably enhance the accuracy and volume of important information on the situation in outer space and improve space traffic management, but it would mean that countries would have to disclose classified and tactical information.

In addition to regulatory and precautionary measures, progress has also been made in the development of debris mitigation technologies. For example, the Dutch Space Agency is conducting a special project aimed at prolonging the life of satellites for up to ten years with the help of a space tug. The ConeXpress Orbital Life Extension Vehicle (CX-OLEV) can ensure the continued use of satellites and prevent their drift as well as any potential collision with other spacecraft.[146] The space tug could also be used for rescue missions. Space debris mitigation technologies like the CX-OLEV could improve space security by reducing the amount of space debris, minimizing the chance of collision between space debris and functional satellites and decreasing the number of inactive satellites in orbit. On another front, NASA's *Implementation Plan for International Space Station Continuing Flight* outlines its goal of installing several protective shields in order to prevent damage caused by space debris.[147] China has established the Space Target and Debris Observation and Research Centre at the Chinese Academy of Science. The centre will research and closely monitor space debris and establish risk management systems. It has the capacity to track space debris in real time and to identify new debris.[148]

Conclusions

International actors must engage in cooperative dynamics and seek common solutions to the complex and interrelated space security issues they all face today. While the need to protect space has been acknowledged worldwide, how to do it seems to be less clear. The extensive exploitation of space opportunities by state and non-state actors has led to the emergence of new threats in space and posed challenges to space security. Space debris, the increased possibility of collision and frequency interference, space weather and the weaponization of space can all negatively affect space power capacities and, more importantly, threaten security on Earth. As the concept of national security has expanded in recent years, the '*multi-sum security principle*' can account for the multidimensional nature of global security – its environmental, national, human, transnational and transcultural dimensions. Given the importance of space to the larger, global security debate, a state can only advance its long-term security needs by also helping others to enhance their safety and well-being.[149] In other words, in order to preserve international stability, world leaders must link the two issues of avoiding unintentional and intentional harm in order to develop a strong legal regime that enhances the prospects of sustainable space capabilities and safeguarding the space environment for future generations.

Existing international agreements and norms do not provide tailored responses to global challenges and are not adequate for ensuring the security and sustainability of space. However, overcoming national strategic interests to engage in genuine international dialogue on global security is not an easy task for states. First, it requires an excellent understanding of the current space environment and its possible evolution and trends. In other words, policymakers need to gauge their own capacities in comparison with the position of other international actors. The framework of *meta*-geopolitics can facilitate space power analysis and ensure that policymakers look at all the areas in which potential instabilities and threats to space security could arise. As explained in *Neo-Statecraft and Meta-Geopolitics*, 'any geostrategy that is based on *meta*-geopolitics will account for the diffuse and shifting nature of today's security threats. Focusing a country's geostrategy on one single threat will leave a country dangerously unsafe, as doing so will make the country ignore other threats that might come up or "miss the dangerous connections between these threats"'.[150] Chapter 5 attempts such a multidimensional analysis. Through the lens of *meta*-geopolitics I investigate the national space policies and resulting power relationships of four contemporary space-faring nations. In order to provide an all-encompassing picture of the existing space geopolitical environment, I also include the non-state sector and analyse its growing role in the changing space context.

5
Space Power and Applied *Meta*-Geopolitics

This chapter assesses the space power capacities of contemporary leading space-faring nations. The space policies of the United Sates, the European Union (EU), Russia and China are tested against the seven capacities of the *meta*-geopolitics framework, and their respective strengths and weaknesses are highlighted and discussed. This multidimensional analysis also includes a section on non-state actors and assesses the space power capacities of two selected commercial companies – EADS-Astrium and Boeing. A clear picture of geopolitical space dynamics emerges and serves as a basis for assessing space power imperatives, drawing conclusions, identifying governance paradigms and making policy recommendations.

A new space environment

The global geopolitical environment has profoundly changed since the end of the Cold War. From a 'bipolar' balance, driven by the competition between the United States and the Soviet Union, space has moved into a multipolar realm, with an increasing number of new actors and technologies entering this domain.[1] Whereas in the past only a few nations had access to space, today actors from around the globe, including developing countries and non-state actors, compete for access and the utilization of outer space.[2] In order to structure this growing space activity, states are establishing national space agencies, thereby contributing to the institutionalization of space matters.[3] In sum, as defined by Peter, we are witnessing the *internationalization*, *globalization* and *institutionalization* of space.[4]

Space offers numerous opportunities and has become a critical component of the geopolitical posture of global actors. While the United States and the EU together with China and Russia remain the leading states in space, new countries and private actors have recently entered this profitable sector. Not only has the number of space assets grown, but the pace at which they are manufactured and placed in outer space has dramatically accelerated.[5] As space grows to be an important source of geopolitical influence, it is becoming more *competitive*, *congested* and *contested*, making operations in space more hazardous for all its users.[6]

Although the need for enhanced international cooperation and better regulation in space has been extensively acknowledged by all actors, they continue to consider space as a highly strategic domain and its importance for achieving national goals has been constantly reaffirmed. In January 2011, the US Department of Defense (DoD), jointly with the Office of the Director of National Intelligence (ODNI) stated that '[s]pace is vital to US national security and our ability to understand emerging threats, project power globally, conduct operations, support diplomatic efforts, and enable global economic viability'.[7] Yet, might the preservation of state's strategic assets and interests in space be better achieved through international cooperation, instead of defensive and unilateral actions? The answer lies in clearly identifying all the variables and accurately defining the geopolitical structure to be addressed. This is precisely what this chapter attempts to do.

Meta-geopolitics: Space power analysis

I apply the framework of *meta*-geopolitics to analyse the space power capacities of the four leading state actors in space listed above. Although space is still dominated by states, they no longer have a monopoly on space activities. The commercial sector plays an increasingly important role in international space matters. The space industry is a driver of research and development (R&D). It has an enormous turnover and represents a core component of the global economy. In order to provide an all-encompassing picture of the current geopolitical space environment, I include two leading space companies in the analysis: EADS-Astrium and Boeing. Based on these two case studies, I draw general conclusions about the private sector in space.

I concentrate on the analysis of national space policies. As pointed out by Peter, space power cannot be reduced solely to technological supremacy but '[i]t is necessary that a country demonstrates some "political will" and develops the attendant national strategy and policy to exploit the elements of space power'.[8] Therefore, a detailed investigation of national space policies will help identify actors' principles and strategic objectives in space, while allowing projections about possible future geopolitical trends. Based on the outcomes of this comprehensive space power analysis, I propose a set of imperatives for each actor. These imperatives are mine and may not necessarily coincide with existing imperatives as each actor sees them, although in my view they are the most useful in advancing stability and security in outer space while allowing global actors to compete in a non-conflictual environment and thus sustain their space power capacities and national interests. This *meta*-geopolitics analysis can be better understood using the table format presented below. The left-hand column lists the seven power capacities of the *meta*-geopolitics framework and the right-hand columns summarize the space power capabilities and related imperatives.

By considering the key capacities and challenges of some of the leading space actors, a realistic picture of current and future space geopolitical dynamics emerges, thus clarifying the most vulnerable areas of space security and global stability so that they can be better addressed. Predictions, of course, are never certain and some of the developments outlined may prove to have less or more of an impact on geopolitics than others.[9] Nevertheless, the *meta*-geopolitics framework enables us to identify the broad parameters and key variables of potential change and can help statesmen take early and informed strategic decisions that will allow a safer and more sustainable use of space.

The United States: Space power analysis

On 28 June 2010, the White House released the US National Space Policy (NSP). Although the Obama Administration seems to have moved from the unilateral positions of the former administration to be more 'outward looking and inclusive, with a far greater emphasis on cooperation',[10] the concrete actions and decisions of his administration remain to be seen. Nevertheless, at this early stage, most analysts agree that while 'less bellicose' and 'less US-centric' in its wording, the new policy seems to build on the long-standing US nationalistic doctrine of space.[11] Table 5.1 summarizes the seven US space power capabilities and weaknesses, highlighting their respective strategic importance within the NSP. This multidimensional analysis draws a comprehensive and detailed picture of where the United States stands at present and provides insights into how it plans to reorganize and reprioritize its space activities in the future. Based on the outcomes of this space power assessment, I propose a number of imperatives which, I believe, would enable the United States to maintain its supremacy in space in a context of budget pressure and growing competition.

Table 5.1 The United States

Issue area	Space power capabilities	Space power imperatives
Social and health issues	The NSP acknowledges the benefits of space for civilian operations such as advanced medicine, information and communications, but considers them as a positive externality derived from the achievement of other, priority objectives.	In a context of economic and social difficulties, a more balanced approach to the commercial, civil and military uses of space would better serve US internal stability and social justice objectives.
Domestic politics	The Obama Administration has cancelled NASA human space flight missions and outsourced transport to the ISS to the commercial sector.[12] Public interest in space has waned.[13] Internal politics and congressional constrains on space operations. National pride remains an important element of the NSP, to be achieved through technological superiority and diplomatic leadership.[14]	Ensuring the security and sustainability of US space objectives and operations is the best choice to guarantee a return on investments and justify US strategic choices in space. This might help rekindle public enthusiasm and interest in space activities, while avoiding expensive and unpopular missions.
Economics	Key position of the space industry in the national economy. Tight links between space industry, government and military. An internationally competitive national space industry is a high priority of the NSP.[15]	Clarify the position on space-related export control reforms and modify the regime to remove the barriers to commercial and civilian space development.[16] Support the development of the commercial space industry, in particular the most vulnerable sectors. Increase cooperation, both internationally and with the private sector, to leverage resources and reduce the risks of space operations.[17]
The environment	Space technology for climate change monitoring, terrestrial and near-Earth space weather forecasting, natural resources management, disaster	Supporting international initiatives and increasing the amount of information shared will enhance US capacities to address climate change issues, mitigate

	response and recovery support is considered critical to US security.[18]	space weather risks and improve responsiveness to natural catastrophes.
Science and human potential	US leading position in space exploration, scientific research and technology is challenged by its competitors. Space industry faces demographic challenges as a significant number of space experts are approaching retirement. The development of human potential and technical knowledge is a high priority of the NSP.[19]	Address the demographic challenge and ensure that critical space knowledge is handed on to the next generation.[20] A balanced approach of commercial, civil and military uses of space will benefit the development of the scientific and commercial sectors.[21] Stable investment in research and development.
Military and security issues	Centrality of space to US national security. World's most advanced military capacities. Military reliance on space-based systems and high vulnerability to intentional and unintentional disturbances. Vital interest in ensuring the sustainability and safety of space assets and operations.	It is in the best interests of the United States to refrain from conducting ASAT tests and from developing space-based weapons; emphasize multilateral rather than unilateral action; encourage confidence-building measures; support international initiatives to prevent the weaponization of space; reduce the vulnerability of its space systems through enhanced recovery capacity, multiplication of back-up systems and improved space surveillance systems (enhanced cooperation and sharing of information).[22]
International diplomacy	Willingness to consider and discuss the issue of space weapons. NSP affirms US leadership in the establishment of international laws and regulations for peaceful, cooperative and sustainable use of space.	Genuine diplomatic engagement and leadership in the development of a proposal preventing outer space weaponization will enhance US prestige.

One of the turning points of the 2010 US NSP is the emphasis put on international cooperation in space matters. Yet, while the new administration stresses the importance of international dialogue as well as the peaceful and sustainable use of space, it clearly states that free access to space, US national interests and security considerations will be preserved first. Indeed, the growing reliance of the US military and economy on space-based systems has led its policymakers to consider space assets as vital to national security. As is formulated in one of the core principles of the NSP, '[t]he United States will employ a variety of measures to help assure the use of space for all responsible parties, and, consistent with the inherent right of self-defense, deter others from interference and attack, defend our space systems and contribute to the defense of allied space systems, and, if deterrence fails, defeat efforts to attack them'.[23] Given the significance of space assets to US national military power, the new policy also stresses the importance of assuring and sustaining 'future reliable and efficient access to space'[24] and therefore places the focus on the development of sophisticated, independent launch capacities.

In the field of the militarization and possible weaponization of space, consideration of arms control regulation is a significant change of direction in US policy, which has previously rejected such initiatives. In this respect, the Obama Administration gives more importance to international regulation measures and agrees to 'consider proposals and concepts for arms control measures if they are equitable, effectively verifiable, and enhance the national security of the United States and the allies'.[25] This statement does not mean that the United States will immediately commit to binding international agreements, such as the space weapons ban treaty supported by Russia and China. Nevertheless, this less unilateral position seems to offer a better chance for the emergence of *rules of the road* and, thus, the reinforcement of transparency and confidence-building measures (TCBMs) in space.[26]

Another expected area of renewed attention in the Obama space policy is the revitalization of the US commercial sector and national space industry. The heavy reliance of the US military and economy on space assets has often led policymakers to pursue more unilateral stances in space. However, the supremacy of US military interests that resulted in the imposition of strict export controls on space technology in the late 1990s seriously weakened the US commercial sector, particularly the satellite industry, and eroded its competitiveness in the global space market.[27] The US Congress, in order to prevent the export and spread of critical technologies to its rivals, in particular

China, has put in place regulations that treat each component of a civil spacecraft as a weapon. These rules have created difficulties for foreign space companies, making it too complicated to import and creating strong incentives for non-US space companies to use European technology instead.[28] The most advanced remote sensing satellites are now built by firms in Italy (Cosmo-SkyMed), Germany (SAR-Lupe) and the UK (TerraSAR-X).[29] Moreover, these protection measures have also affected the US human capital capacity of *meta*-geopolitics. As reported by the academic space research community, 'highly talented foreign scientists seek positions elsewhere (such as in Russia, China and India), where the restrictions on collaboration are less burdensome'.[30] A lack of commercial investment makes the US space industry even more dependent on military contracts, potentially hampering advances in R&D of civil technology, which could advance even further Europe's position as a global space power.

In parallel, the arrival of new state and private actors in the space business reduced the US share of global satellite exports by almost two-third in 2008.[31] As explained by Gregory L. Schulte, 'the US competitive advantage in space has decreased as market-entry barriers have lowered, and the US technological lead is eroding in several areas as expertise among other nations increases'.[32] In such a context it was not a surprise that the rehabilitation of domestic space industries and commercial space have been placed at the top of national priorities by the Obama Administration. A stable space industrial base will drive down the costs of space technology, generate important revenues, create additional jobs, improve US international competitiveness and foster technological development. This trend has been reflected in the US budget, with the President cancelling NASA human space flight missions, including the Bush-initiated Constellation, which was supposed to send astronauts back to the Moon and then to Mars. NASA has been offered an additional budget of USD 6 billion over the next five years, mostly to support the development of commercial flights of crew and cargo to the International Space Station (ISS).[33] Outsourcing NASA's routine space transportation activities should free up its limited financial resources for complex R&D to enable the United States to further explore the solar system.[34]

Although the policy is not very clear about space-related export control reform, it states that improvements in the commercial space sector cannot be achieved without government support and a pool of highly skilled space professionals.[35] In order to sustain its leadership position in space, the United States needs to build up a pool of highly skilled experts and invest in scientific R&D, '[t]his will entail enhancing recruiting, retention, and training policies designed to develop current and future national security space professionals in the military, intelligence, civilian, and contractor components of the workforce.'[36]

Given the fact that a considerable number of space experts are approaching retirement, the Government must take appropriate measures to ensure the transfer of know-how and expertise to the younger generation of engineers and meet the needs of a growing space industry.[37] Given the tight relationship between the different space power capacities, investment in innovative space technology will, in turn, have positive spin-offs such as better understanding of climate change and the space environment, the emergence of international and private sector partnerships, the creation of additional jobs nationwide, the strengthening of US prestige and, hence, the reinforcement of its leadership role on the international stage. The Obama Administration clearly positions the United States as a leader in the domain of space and commits to '[d]emonstrate US leadership in space-related fora and activities'.[38] The NSP stresses the responsibility that the United States has as the dominant actor in space, the role of which is to lead and foster multilateral efforts for securing the utilization of space.

The multidimensional analysis of US space power capacities and space policy carried out in this section highlights interesting power relationships and geopolitical trends. On the one hand, the United States calls for more international cooperation to ensure the responsible, peaceful and sustainable utilization of space by all actors. Such cooperation is particularly relevant in the fields of space exploration, space surveillance and Earth observation.[39] On the other hand, it remains heavily influenced by military and nation-centred visions of space, as formulated by its predecessors. To maintain freedom of access to space, strengthen economic interests and the US reputation, create a highly skilled space workforce and, of course, defend US assets and interests in space remain the top priorities of US space policy. The United States obviously continues to consider space crucial for its vital national interests and to prioritize military and commercial uses of space over social applications.

One of the reasons for this is to be sought in the heavy reliance of the US military and economy on space-based systems and the tight links between the US government, commerce and the global space industry. During the early years of space exploration and technology development, space was primarily viewed as a crucial component of Cold War era national security competition. As such, space research and activities were primarily government-funded and organized by the military. The recent shift by the global space industry to more commercial applications, however, means that the relationship between government, business and space R&D has evolved into a complex configuration, full of overlapping tasks and interests. Today, much of the US commercial space industry's demand comes from government agencies, and it is important for the future viability of the space industry that government policies continue to stimulate demand for space-based technologies.[40] The US Government also has a good reason to support the commercial space industry because many of the technologies can be produced more efficiently and at a lower cost by private companies. Furthermore, US military and national security operations increasingly rely on space technology, including space commercial systems. As explained by Richard Weitz,

> [t]he Pentagon uses space-based assets for communications, reconnaissance, navigation, targeting, and other core military activities, while satellites provide vital information to the US intelligence community. These assets are essential for important US national security missions including conducting combat operations, verifying arms control agreements, analyzing foreign defense developments, and monitoring long-term environmental conditions. The Pentagon's commercial, civil, and foreign partners also rely on unfettered access to space for economic, scientific, and international missions that benefit the United States. The Pentagon relies on commercial satellites for much of its communications and imagery, especially in Afghanistan.[41]

Thus, the US military has become highly dependent on commercial satellites for its communications and reconnaissance missions. As a result, commercial US space assets can be directly linked to US security.[42] The fact that most space technology is dual use by nature makes the military even more dependent on commercial space systems and private sector innovations.[43]

Its extensive presence in space has made the United States highly vulnerable to unintentional harm through such incidents as collisions, space debris or severe space weather events as well as potential intentional attacks by its adversaries. The DoD and ODNI, in their 2011 joint *National Security Space Strategy*, expressed their concern that as other countries and non-state actors are developing advanced space capacities, and perhaps gaining relative advantages, space is becoming 'congested', 'competitive' and 'contested'.[44] These three threats are considered the most serious challenges to US national interests and supremacy in space. This supremacy is somewhat ambiguous, it is the source of US geopolitical power but also makes it the most vulnerable of all the players in space. Although Russia and China cannot compete with the United States in missile defence or nuclear and conventional weapons, they do have the ability to engage in asymmetric responses.[45] Adversaries increasingly utilize techniques such as electronic and communications jamming or cyber attacks to counter US space capabilities. In addition to these intentional threats, a collision with debris or another satellite could also disrupt US space-based systems and seriously affect its space power capacities. Events, such as the 2007 Chinese ASAT test, 'not only re-emphasized the vulnerability of [US] satellites to direct attack, but the resulting debris imperils everyone's satellites'.[46] The 2009 unintentional collision between US and Russian satellites 'brought space debris and space situational awareness to the forefront of concern for everyone who operates satellites, both governments and commercial companies'.[47]

The Obama Administration takes these issues very seriously. The National Space Policy and the National Security Space Strategy both outline a US action plan to prevent and deter aggression against the United States and its space-based vital systems, in the 'contested' space environment of today. While the continuation of monitoring efforts such as Space Situational Awareness (SSA) systems or Space Debris Mitigation (SDM) practices remain an important component of this approach, the US administration calls for more collaboration in further development of such systems, as well as responsible behaviour and enhanced sharing of information regarding situation awareness in space. According to William J. Lynn, '[f]urther expanding the amount and kind of data we share will, over time, help foster the sustainable space environment that our own strategic advantage depends on'.[48] Another important deterrent measure presented in the US strategy is the creation of coalitions in space, so an attack on one of the members will be perceived as an aggression against them all.[49] In addition to its deterrent function, such alliances will allow cost-savings and increase operational capabilities. Making US space-based systems more resilient and its military power less reliant on them are further objectives of the new administration.[50] However, the former would have a limited impact, as satellite vulnerability can only be mitigated but in no way eliminated.[51]

Moreover, the President has expressed his commitment to reinforce US leadership in order to foster international dialogue in space security matters.[52] The unilateral responses pushed by former President

G. W. Bush to address space security issues proved counterproductive to US national interests.[53] A good example is given by the Union of Concerned Scientists: '[w]ithout constraints on ASAT weapons, for example, threats to satellites will continue to proliferate and mature, requiring the United States to expend more effort securing satellites and leading to less predictability and stability in crises'.[54] Space security is essential to assure and enhance US capabilities in space but it seems to be difficult to achieve through militarization and unilateral defensive measures. Instead, as is rightly pointed out by Schulte, '[r]ules can help the United States minimize the chance of collisions in space, reduce unintentional radio frequency interference, maximize the use of crowded orbits, and discourage destabilizing behavior such as intentional interference with space systems in times of crisis. Rules encourage good conduct but also provide a way to hold accountable those who would engage in malign acts.'[55]

For all these reasons, there is a strong case to be made for the United States to enhance its cooperation in space and play a leading role in setting standards and norms for the peaceful and sustainable use of space by all actors. Increased cooperation would help the United States to leverage resources and share the risks inherent in space operations, enhance US diplomatic prestige and improve its political sustainability and workforce stability.[56] Yet, despite these benefits, a certain number of obstacles continue to prevent the development of additional US cooperation initiatives. Existing internal politics and congressional limitations hamper the cooperative engagements of the Obama Administration. As Marcia S. Smith points it out, '[f]or the USA to have a space program at all it must have support from taxpayers and their representatives in Congress and there are many elements of the Obama policy that affect primarily US domestic activities'.[57]

One of the most controversial debates concerns the proposed partnership between NASA and the private sector. It raises questions over whether a commercial venture can be funded by taxpayers and whether the private sector would be able to meet crew safety requirements. There are also congressional limitations on US cooperation with China.[58] Congress formally opposes such cooperation and section 1340 of the FY2011 Continuing Resolution prevents NASA or the Office of Science and Technology Policy from utilizing any funds to 'develop, design, plan, promulgate, implement, or execute a bilateral policy, program, order, or contract of any kind to participate, collaborate, or coordinate bilaterally in any way with China or any Chinese-owned company',[59] unless expressly authorized by the Congress. This provision also prevents

NASA from using any funds 'to effectuate the hosting of official Chinese visitors at facilities belonging to or utilized by the National Aeronautics and Space Administration'.[60]

The technology transfer constraint is another obstacle to increased US international cooperation and needs to be reviewed and amended by the US administration and the US Congress.[61] Finally, the narrow national security view of space remains a core element of US space policy and while it allows the development of soft law tools for arms control in space, it does not give any more chance than past documents to the negotiation of a legally binding international space framework.

Looking at US space power capacities and National Space Policy from a *meta*-geopolitics perspective provides an all-encompassing picture of US strengths and weaknesses in the changing space environment, with the new opportunities and threats it presents to US space power. US superiority in terms of space assets ranging from satellites, global positioning systems and remote sensing technologies, coupled with its extensive strategic utilization of space technology in the seven areas of *meta*-geopolitics, undeniably confirms the leading position of the United States in space. However, whether the United States will remain the leading space power in 'a more global space landscape with growing strength in emerging economies and increasing specialisation in various parts of the world'[62] remains to be seen. The *meta*-geopolitics framework helps us identify existing and potential vulnerabilities and predicts the possible evolution of US strategic objectives and behaviour in space. The analysis of its space power capacities highlights that the United States has a clear self-interest in preserving the stability and sustainability of the space environment. However, it cannot address these complex space security issues unilaterally, and its leadership is essential to overcoming existing deadlocks.[63] Yet, even if US space policy sounds more 'cooperation-oriented' in its wording, as rightly observed by Jeff Kueter, the President of the George C. Marshall Institute, '[b]y itself, the policy is just a statement'.[64] The way this statement is translated into concrete action and positions within international cooperation forums remains to be seen and will determine its genuine and long-lasting impact on US national space power and future geopolitics.[65]

The European Union: Space power analysis

The EU has 27 member states and, together with Norway and Switzerland, both members of the European Space Agency (ESA), 29 European countries cooperate in space. The European space programme is vital evidence of the benefits of cooperation in space, since each individual country is too small to emerge as an international space power with a well funded and advanced space programme. By joining forces through cooperation efforts, the European countries have managed to become one of the most powerful actors in space. Together, they have an advanced launch capability and a satellite manufacturing industry and are engaged in advanced and progressive space research. Europe has a well developed civilian and commercial space sector and its extensive experience in international cooperation makes it a strong and important actor in the international space landscape. Table 5.2 highlights the EU's strengths and weaknesses and the imperatives that would better enable it to enhance and maintain its space power status in an increasingly competitive space environment.

In recent years, the EU has been pursuing deeper integration and expansion of its membership, while trying to maintain and strengthen the coherence of its common action. One of the notable results of these continuing efforts was the signing, in 2007, of the Lisbon Treaty, which entered into force on 1 December 2009.[77] Article 189 of the Treaty is of particular interest as it deals with space matters and provides the EU with a robust legal mandate in the field of space:

1. To promote scientific and technical progress, industrial competitiveness and the implementation of its policies, the Union shall draw up a European space policy. To this end, it may promote joint initiatives, support research and technological development and coordinate the efforts needed for the exploration and exploitation of space.
2. To contribute to attaining the objectives referred to in paragraph 1, the European Parliament and the Council, acting in accordance with the ordinary legislative procedure, shall establish the necessary measures, which may take the form of a European space programme, excluding any harmonisation of the laws and regulations of the Member States.
3. The Union shall establish any appropriate relations with the European Space Agency.[78]

For the purposes of the present analysis, I consider the EU to be a single supra-state actor and focus on the EU's common activities in space, instead of investigating the position of individual member states. With the multiplication of actors and globalization of the space environment, Europe is facing new challenges and constantly evolving threats to its security that have become more diffuse, and more difficult to identify and predict than before. While the EU has the

Table 5.2 The European Union

Issue area	Space power capabilities	Space power imperatives
Social and health issues	The European Space Policy highlights the importance of space to best exploit its social opportunities and improve citizens' quality of life.[66]	The EU has a long-standing and developed civilian space programme whose services can be improved through better integrated and coordinated programmes.
Domestic politics	In the EU case *integration* is more relevant than *domestic politics*. The European Space Programme can be seen as a path to strengthen European unity.[67] Space exploration remains an ideal example of EU cooperation and a central element of EU prestige on the international stage.[68]	Improve coordination between the ESA and the EU, and among member states in the field of space. Set up clear responsibilities for the realization of core space programmes with realistic budgets. Find new sources of funding for EU space programmes as ESA member states are facing important budget cuts.[69]
Economics	The European space industry represents important shares of the global space market and is therefore critical to the EU economy. A competitive European space industry is part of the EU's high strategic priorities.[70]	Advance revenue- and job-generation programmes such as Galileo and Global Monitoring for Environment and Security (GMES). Support private-public ventures to provide sustainable funding for EU space projects and foster the competitiveness of the EU's space industry. Further increase international cooperation to leverage resources and reduce the risks of space activities.[71]

The environment	Earth observation-based services already exist in Europe but at the national or regional levels. Environmental issues and the GMES programme are a high priority of the new EU Space Policy. The GMES initiative demonstrates the EU's political will to affirm its leadership in the global fight against climate change.[72]	Advance the EU-ESA jointly launched initiative on GMES/Kopernikus.[73] Allocate sustainable funds and resolve governance issues hampering the progress of the GMES programme. Supporting international initiatives and increasing the amount of information shared will enhance EU capacities to address climate change issues, mitigate space weather risks and improve responsiveness to natural catastrophes.
Science and human potential	The development of technical expertise and a qualified workforce has been at the heart of EU space cooperation. Technological knowledge will enhance the EU's global competitiveness and prestige. Advances in space technology and investment in human potential remain among the top priorities of the EU.	Support private-public projects that will boost innovation and R&D, particularly in high-technology and high-capability industries.
Military and security issues	The EU has been slow to develop its military space capabilities critical to dealing with today's global security issues.[74] Galileo and the GMES system are designated priority programmes, although a consensus is difficult to find on their security applications.	Overcome the political deadlock and encourage progress in the Common Foreign and Security Policy/Common Security and Defense Policy (CFSP/CSDP). Allocate sustainable funds, provide genuine political support and resolve governance issues to advance the Galileo and GMES programmes.

(*continued*)

Table 5.2 Continued

Issue area	Space power capabilities	Space power imperatives
	EU space military and security dimensions are hampered by national sovereignty considerations and budget constraints.	Refrain from conducting ASAT tests and from developing space-based weapons. Ensure interoperability of EU space systems with existing ones to reduce frequency interference risks. Support international efforts aimed at preventing the weaponization of space. Advocate for the EU Code of Conduct and support other confidence-building initiatives.
International diplomacy	The EU has high visibility in international forums such as COPUS, the ITU and the CD.[75] In February 2009 the EU launched a draft Code of Conduct on outer space, reflecting a common EU position. It was submitted for consultation during a meeting of the CD and also for discussion with non-EU space-faring nations.[76]	Speak with a common voice and take clear positions on critical space issues such as the weaponization of space. Advance the GMES programme to reinforce EU leadership in global climate change matters.

potential to maintain its space power position in this changing structure, to be successful, it has to consider the new variables and adapt its strategic goals and actions in space.[79] In order to reflect the recent developments in the EU integration process provided by the Lisbon Treaty and better address the changing geopolitical space context, the European Commission issued in April 2011 a communication laying the basis of a new integrated European Space Policy.[80]

As is indicated in the policy's title, *Towards a Space Strategy for the European Union that Benefits its Citizens*, the Commission first highlights the societal and economic benefits of space, such as EU independence and competitiveness, improved quality of life for its citizens as well as a reinforced position in the international space dialogue.[81] These goals are to be achieved through reinforced and better protected space infrastructure, and technological advances combined with spin-offs in other industrial sectors as well as enhanced cooperation among EU member states, the ESA and the other space-faring nations – the United States, Russia and China.[82]

In line with these new strategic priorities, the competitiveness of the European space industry in the global market is given particular attention. In fact, the EU already supplies an important share of global commercial demand for satellites, and launch and communication services.[83] With the global space services market growing at an exponential rate, a vigorous EU space industry would generate important revenues and therefore represents a core element of EU economic power. Given these optimistic business projections, the advancement of the Galileo satellite navigation programme coupled with the European Geostationary Navigation Overlay Service (EGNOS)[84] and the global Earth monitoring programme (GMES)[85] has been placed by the EU at the top of the ESA's priorities. Moreover, EU advances in space have been acknowledged as of critical importance for two flagship initiatives: the European 2020 strategy[86] and the EU Industrial policy initiative.[87] Indeed, a dynamic and healthy space industry would stimulate economic growth and the creation of additional highly skilled jobs in Europe. Furthermore, it will foster innovation and advances in space technology, with benefits cascading to other branches of EU industry, contributing to the EU objectives of technological independence and enhanced global competitiveness.

An independent and competitive EU space industry requires technological knowledge and a highly skilled workforce, relevant to the science and human potential capacity of *meta*-geopolitics. As noted by the Commission, 'it is necessary to support research into critical technologies (i.e. those that are essential for the sector's strategic non-dependence) and breakthrough technologies (i.e. those that constitute genuine technological advances), including research supporting space exploration'.[88] Yet, the advancement of sciences and space technology is not a new goal for Europe. In contrast to the development of the US space programme, which was rooted in its military applications, early European space endeavours were based on scientific and technological cooperation between member states. Through the ESA, which was created in 1975 and aimed to 'provide for and to promote for exclusively peaceful purposes, cooperation among European States in space research and technology and their space applications, with a view to their being used for scientific purposes and for operational space applications systems',[89] Europe has gained significant scientific and technical expertise that has contributed to enhancing the competitiveness of the EU space industry in the world market. The Ariane programme, the first rocket of which was launched in December 1979, was a clear illustration of European technological and commercial success and, more importantly, independence in space.[90]

Furthermore, the way the ESA collaborated with the private sector to carry out its projects also contributed to its technological development and indirectly supported the EU integration process. Thus, instead of distributing contracts to selected companies, the ESA promoted its members' cooperation in space by giving preference to the creation of private multinational networks such as MESH (Matra, ENRO, Saab and British Aerospace) and STAR (British Aerospace, Dornier, AEG-Telefunken and VFM).[91] The organization contracts with European companies on a principle of 'fair return', which means that it applies a ratio between the shares of an ESA member state in the weighted value of contracts and its share of the contributions paid to the agency.[92] The establishment of such consortia allowed the ESA to capitalize on private sector resources, helping it to remain competitive in a highly complex and demanding environment and develop creative, forward-looking technologies.

In its recent communication, the European Commission highlights once again the strategic importance of investing in space research and innovation to enhance the worldwide recognition of 'European research excellence'[93] and lead to technological self-reliance. The Commission also points out the close interrelationship between the various dimensions of space power and stresses that technological success in space will contribute to diversifying the range of possible terrestrial applications and thereby enhance EU industrial competitiveness, foster economic growth and employment while improving general living conditions for EU citizens. Besides the Galileo and GMES programmes, space exploration has been advanced by the Commission as another way of 'engaging in some of the highest technological challenges of today and tomorrow, ensuring Europe of an ever increasing innovation and know-how'.[94]

In addition to the economic and technological benefits, space has played an important role in the European integration process and remains a critical element of what I define as EU domestic political capacity. Indeed, the high level of cooperation between member states and the harmonization of national space policies achieved in the framework of the ESA made a significant contribution to European integration.[95] As explained by Sheehan, the creation of the ESA in the early 1970s illustrated 'that European space science and technology had significantly matured and with a decade of cooperative experience within the space organisations[96] and within the EEC itself, the political views of the European states had also matured and Europe was now capable of acting as an entity with a clear political will and consistent policy'.[97]

In order to understand the past and future positions of Europe in space we must take into consideration European specificities. The EU is a supra-national actor that is pursuing deeper integration and expanding its number of member states. In space matters, like in any other political domain, this could make agreements between EU member states more difficult and render the policy documents and official positions weaker. In order to gain and maintain weight as a global space actor, the EU must be perceived as a solid unit, and member states must be ready to engage in a robust European space strategy. It is not only the EU that suffers from a complex decision-making process, but coordination between the member states of the ESA is another potential complication to take into consideration. The coordination efforts between the EU and the ESA might also be a source of disputes. In order to address these challenges, while acknowledging the existing differences between member states in terms of political commitment, and financial and technical capacities, the Commission has called for more efficient space governance, enhanced cooperation between member states within Europe and the ESA as well as improved collaboration between the two organizations.[98] It stresses the political dimension of space and advanced space exploration as one of the activities capable of 'merging the interests of the different Member States'.[99] For example, increased participation by the EU in the ISS, making sure that all member states take an active part in it, could help foster cooperation between European nations, inspire a feeling of European pride and encourage member states to behave as a single actor in the fast-changing and globalized space landscape.[100]

Although early European efforts in space were led through the ESA and were constitutionally limited to the peaceful uses of space, today Europe, like other space-faring nations, clearly asserts its space-based security and defence interests.[101] This shift in the EU's vision of space was the result of a combination of factors. On the one hand, the end of the Cold War, coupled with accelerated European integration and advances in space technology, extended the scope of EU competences to the domain of space. As explained by Sheehan, as more areas of human life were increasingly affected by space technology, the need grew for larger EU responsibilities in space matters.[102] On the other hand, the changes in the security environment resulted in a broadening of the definition of global security itself and the gaps in the initial ESA definition of the 'peaceful' utilization of space facilitated the extension of the ESA mandate to include the utilization of space assets to ensure the security of EU citizens.[103] Security is to be understood here within its multidimensional perspective, discussed in Chapter 4, by applying the '*multi-sum security principle*', which identifies five dimensions of global security (human, environmental, national, transnational and transcultural).[104] This definition of global security is much broader and more complex, spreading beyond a traditional military understanding that focuses on economic, environmental and human threats to society. Such an extension of the concept of security enables the ESA and Europe to endorse new roles, and for the latter to 'assert itself as a legitimate "security" actor. Its core competences, such as economic and environmental coordination, could now be presented as "security" issues. Having established a legitimate presence in the debate on "security" it was then only a small step to asserting competence in unambiguously military issues.'[105]

Thus, recently gained 'security' space power capacity has been acknowledged by the EU in different official documents. One of them is the *White Paper on European Security* issued in November 2003, where the Commission clearly states that '[s]pace has a security dimension and security has a space dimension',[106] and highlights the importance of space technology to the EU's Common Foreign and Security Policy (CFSP) and particularly the Common Security and Defence Policy (CSDP).[107] The European Defence Agency (EDA) was established in July 2004 to help member states improve and implement the CSDP.[108] Although the EDA does not have a set of formalized provisions for military space capabilities, its *Long-Term Vision* document emphasizes the role of space technology as a primary means of information and intelligence gathering. The agency concludes that missions under the CSDP will depend on the capacity to gather, share and disseminate such information gained through space capabilities and therefore that the development of such capabilities is vital to the success of CSDP operations.[109]

It is important to note that many European states are also members of the North Atlantic Treaty Organization (NATO) and have submitted to its policies and contributed to its capabilities. NATO does not have a specified policy, committee or working group to guide its space operations,[110] but it is currently making use of space capabilities both as space control and in-force enhancement. NATO's space capabilities consist of an advanced satellite communications system (SATCOM), missile warning and missile defence, weather information, SSA and space-based intelligence gathering. NATO also uses the US GPS to provide positioning, navigation and time data for its forces. It has not yet been decided what level of use NATO will make of the Galileo system once it is in full operation.[111]

In May 2007, the European Union Space Council and the European Commission published the first comprehensive *European Space Policy*, which reaffirms the relationship between space and the successful implementation of the CSDP and discusses pressing European security issues.[112] The policy statement analyses the challenges presented by the dual-use capabilities of space technology and calls on the EU to focus on considerably improving coordination between its civilian space and military/defence programmes.[113] Yet, this document does not address the implicit global challenges that arise from Chinese and US testing of ASATs or the US space doctrine and, for this reason, it has been argued that the European policy 'is somehow reluctant to take responsibility in the framework of international and global security'.[114] Among its most notable shortcomings, the EU Space Policy published in 2007 does not address the possible danger of an arms race in outer space. As such, it has been argued that the EU missed an opportunity to be in the vanguard of space policy by deliberately ignoring the global context of space issues – civil or military.[115] It is however important to note that the EU consists of 27 individual states, and that the highly sensitive issues of arms control and security policy relate to individual national security policies. These are areas where it is difficult to get agreement on the appropriate policy between all member states, and this is one of the reasons that there was no explicit mention of such controversial issues in that document.

The recent Commission communication on the EU's new space policy acknowledges the 'European dimension of space for security and defence' and presents security as a critical area of future European space efforts, with Galileo and GMES as leading projects to enable Europe to meet its global security goals.[116] It also cites the Space Council, which reaffirmed at its seventh meeting the security dimension of space policy at the service of the EU's CSDP and reminded Europe of its duty 'to explore ways to support current and future capability needs for crisis management through cost-effective access to robust, secure and reactive space assets and services [...] taking full advantage of dual-use synergies where appropriate'.[117] Given the political sensitivity of the discussions on EU common security and military capacities, the Commission, once again, adopted a very cautious approach to the EU's military application of space technology.

GMES therefore exclusively revolves around global environmental threats. Developed by a heterogeneous pool of stakeholders, including the EU and the ESA, GMES has been designed to monitor land, ocean, atmospheric and air quality changes.[118] Once fully operational, it will provide Europe with autonomous access to space data and enable informed and independent decision-making on a set of critical environmental issues. Moreover, the GMES system has the potential to reinforce Europe's position in the international dialogue on global climate change and connected issues. The GMES is an ambitious programme from a technological and scientific point of view that intends to tackle the weaknesses of the global Earth observation sector, 'namely meeting the daily needs of users, ensuring data continuity and distributing space-based data in an integrated information system'.[119] It also constitutes an important contribution to the Global Earth Observation System of Systems (GEOSS), the leading international project for Earth observation information gathering. Offering the international community a powerful tool to address global environmental issues will improve the EU's image and visibility on the international stage.[120] Furthermore, like most space-based systems, GMES is inherently a dual-use technology. In its recent communication, the Commission recognizes the GMES security component and its efficiency in ensuring European transnational security, for example, through improved maritime surveillance and border control. However, despite this effort to define the security component of the GMES, its concrete security applications remain controversial because of their highly political nature. Developments in this sensitive field remain closely tied to the overall progress of the CFSP/CSDP.[121]

One of the other major, dual-use projects realized in collaboration between the ESA, the EU and the private sector is the Galileo satellite navigation and positioning system. Given the serious geopolitical implications of satellite navigation systems outlined in Chapter 2, this European initiative once sparked serious controversy between the EU and the United States. The then President of France, Jacques Chirac, in December 2001 warned that without its own satellite navigation system, European countries risked becoming 'vassals' of the United States. He particularly mentioned two main motivations for the Galileo programme: defending European sovereignty and challenging US strategic and technical leadership.[122]

The EU initiated Galileo by arguing that it would be a purely civilian and commercial system, as opposed to the militarily controlled GPS. However, some critics claim that Galileo has been connected to military uses from the beginning by its so-called Public Regulated Service, in which encrypted signals can be used by military and security agencies.[123] The European Parliament passed a resolution on 10 July 2008 recognizing 'the necessity of Galileo for autonomous ESDP operations, for the Common Foreign and Security Policy, for Europe's own security and for the Union's strategic autonomy'.[124] This is a change in position, since the European Parliament had previously rejected all hints of a military role for Galileo. However, the European Parliament also concluded that the European Space Policy should 'under no circumstances [...] contribute to the overall militarisation and weaponisation of space'.[125] Since no single European country opposed the GPS, the Galileo project moved forward, challenging the US monopoly and potentially affecting its navigational dominance. Satellite navigation systems are high-technology and high-capability industries, and add significantly to a country's geopolitical power. The construction of an independent global positioning satellite system is essential for the EU to handle the requirements of today's military operations and to assert itself as a global actor on the international stage without being undermined by its continuing dependency on third party technology and information.[126]

The last security dimension of the new European Space Policy deals not only with the protection of space infrastructure from 'natural phenomena, such as solar radiation and asteroids, and [...] other spacecraft and their debris',[127] but also against intentional threats, with electromagnetic interference given as a concrete example. The Commission states that efficient protection of EU space assets will be guaranteed by the implementation of European SSA and the development of independent monitoring capabilities.[128] The Space Situational Awareness Preparatory Programme (SSA-PP) was approved at the November 2008 ESA Ministerial Council and launched in January 2009, for an initial three-year trial period.[129] Once fully operational, it is planned to be a European version of the US Space-Based Space Surveillance Programme and is intended to create a basis for future political, diplomatic, regulatory and technical measures to guarantee safe access to space. It will provide Europe with an independent system of information on objects in orbit, space weather and other threats, such as asteroids and comets.[130] An incident that shows the importance of this project happened in December 2008, when a European weather satellite came close to collision with a piece of debris from the Chinese ASAT test. The ESA did not anticipate this event and was notified of the potential collision by the US Air Force.[131] This demonstrates Europe's current reliance on the US military for information about space events, something that can be avoided by the successful establishment of an independent SSA programme. Moreover, such a powerful tool will reduce the potential financial costs of the loss of European space-based systems caused by collisions with debris and space weather.[132]

With the rapid transformation of the space landscape it is critical that European policymakers get a 'sound understanding of space power in Europe to be able to develop a strategy that maintains Europe's advantages in the space sector to enable Europe to maintain its position'[133] in the emerging new space order. The *meta*-geopolitics approach applied to the investigation of European space power capacities and its strategic objectives, as formulated in the policy statement by the European Commission in April 2011, allows identification of the major European achievements and the remaining challenges, while giving an insight into the possible future evolution of the EU.

The outcome of the *meta*-geopolitics analysis confirms that Europe, when considered collectively, is active in all the seven fundamental space power capacities and enjoys a comfortable position in the contemporary space geopolitical landscape. Unlike the US space programme, which is largely dominated by the military and national security dimensions, Europe, through the ESA, capitalizes on its economic, technological and scientific advantages in space. However, as European integration moves forward, and with the adoption of the first EU Space policy and the CSDP, Europe has shifted to a more political vision of space and taken the necessary steps to embrace its diplomatic, security and strategic advantages. Interestingly, in the particular case of Europe, 'it is the civil and commercial space industry that has set the stage for military space programs, not the reverse'.[134] In the past two decades, science and technology in Europe have been advancing at a fast pace. The European space industry has been focusing on extending the applications of space to various civilian purposes, with a positive impact on the EU's economy, risk and disaster management, the environment and the health care sector. In recent years, the ESA has increasingly undertaken programmes that have dual-use applications, such as the SSA, the military requirements of which have been provided to the ESA by the EDA.[135] In parallel, EU member states have pursued individual space military endeavours, but the entry into force of the Lisbon Treaty coupled with the development of the CSDP and future national budget limitations are pushing the EU towards more harmonization and improved cooperation in the development of space systems to support military operations.[136] Europe has also gained visibility and influence in international forums and is an active member of the COPUOS, the ITU and the CD. While challenges in space remain, the collaboration between EU member states and other non-European countries in the framework of ESA has made Europe a serious competitor to US hegemony in space.

However, one of the main obstacles that the EU faces in space is the same one that it strives to overcome in many other critical areas: how to increase the scope and depth of cooperation while expanding EU membership. The EU has in past decades moved from comprehensive economic cooperation to an increased focus on political integration. It is difficult, however, to produce and maintain an efficient and coherent space policy that must be adopted unanimously by 27 member states, which have major differences in priorities, financial and technical capabilities and security scenarios. The differences in national positions relating to the highly sensitive military use of space and the allocation of budgets to military activities represent an obstacle to deeper EU integration in space. Furthermore, the fragmented policies and the coexistence of national and ESA programmes sometimes result in overlap, redundancy and inefficient uses of resources. As pointed out by Sheehan, '[w]hile the Europeans' struggled to operationalise their space goals during the 1960s, their difficulties simply mirrored those being experienced in the economic and military dimensions of the European integration project'.[137]

Hence, in order to preserve its space power position, the EU needs to build on the collaborative experience realized in the framework of the ESA to 'shape a coherent and strong European space identity'.[138] It is therefore vital that the EU shows genuine political will to act as a unitary actor in space and develops associated robust space policies and strategies. It must take a common position on critical issues, such as the militarization of space, and play a proactive role in the negotiation and drafting of international rules that will regulate future global space activities. It must work to improve EU space governance and develop fruitful partnerships while strengthening cooperation with the other space-faring nations. Europe must search to enhance its space power capacities and acquire more independence in space, as '[o]nly a strong and independent global space power, possessing unquestionable technical skills, durable infrastructures and an effective and outward-looking organisational structure can expect to meet the challenges of the multipolar order currently emerging, to protect its own interests and strengths, while ensuring that all members of the EU, especially the newest, enjoy the resulting material advantages.'[139] Yet, all these efforts will be ineffective unless properly supported and funded. It is therefore critical that European space endeavours receive financial support from various public constituencies and that EU policymakers allocate adequate resources to space activities.[140]

The European Commission communication of April 2011 is obviously an attempt to address these issues and position Europe as a major space player on the international stage. In the communication document, particular attention has been paid to European independence and technological self-reliance, with the Galileo, GMES and SSA projects pushed to the top of ESA's programme priorities. The Commission addresses defence- and security-related space applications and calls for increased cooperation and harmony between military and civil EU space programmes. Although European military capabilities in space are currently far behind those of the United States, there is a growing realization of the importance of such technological assets and an emerging commitment by the European Commission, the ESA and the EDA to improve the range of space capabilities in Europe, especially in positioning, navigation and time standardization systems. The document also highlights the economic and societal benefits of space and calls for more efficient governance of European space activities and enhanced cooperation with the United States, China and Russia.

In addition to the recently published Commission document, which opens the way to a new European space policy, Europe made another notable step to advance its global space power position by issuing in September 2010 a *Draft Code of Conduct for Outer Space activities*.[141] The EU thus places itself as a mediator and proposes a non-legally binding document, through which states can voluntarily adhere to a certain number of rules and regulations in space.[142] The core principles of the

Code invite all actors to avoid harmful actions in space that could damage or destroy satellites, interfere with their communications or create more debris. More importantly, however, section 4.5 of the Code calls on '[the] Subscribing States [...] to promote further security guarantees within the appropriate fora for the purposes of enhancing the security of outer space activities by all States and the prevention of an arms race in outer space'.[143] Speaking with a common voice, Europe finally managed to put the highly sensitive issue of arms control in space on the EU's agenda. After its publication, the Draft Code was submitted for consultation to the US administration and other nations. The US administration, usually reluctant to engage with any external proposal in the field of space and arms control, expressed its interest in the EU Code, but there was no official approval for the Draft.[144] The Draft Code of Conduct appears to be the most important European diplomatic success in the field of space in the twenty-first century. And it 'exemplifies the EU's current evolution into a more active international player in space matters'.[145]

While the EU has made significant progress in increasing its influence as a global space power, its efforts might be seriously undermined by poor governance mechanisms and financial issues that remained unresolved.[146] Discussions about the ESA's future as a separate organization, with its own members and budget, or as a 'formal arm' of the European Commission are still ongoing and this lack of clarity might hamper further cooperation between Europe and the ESA.[147] The European Commission rightly stresses the strategic importance of Galileo and the GMES programme, but does not address the financial difficulties of these large-scale projects.[148] As regards the funding of the new space policy, the Commission's position is also vague.[149] Yet, given the foreseeable cuts in future EU budgets, it is likely that funds allocated to the EU space programme will not meet the EU's space ambitions. The presentation of the European space programme is supposed to provide more detail on the funding of EU Space Policy and will therefore be critical for future European influence in space.

China: Space power analysis

China's space policy and its space programme are closely linked to the country's Five-year Plan for Economic and Social Development, which was issued in 2006.[150] Since the publication of a White Paper outlining China's space goals, principles and achievements for the next five years, no recent strategic space document has been issued. However, in March 2011 the People's Congress endorsed the 12th Five-year Plan for National Economic and Social Development, which sets out the strategic goals for China for the next five years. Particular focus is directed to scientific development, economic growth, robust industry and services sectors, enhanced regional and international cooperation, environmental issues and improvement of social welfare.[151] Given the importance of space in helping China achieve these objectives, the recently published plan could influence China's future strategic action in space. Table 5.3 summarizes China's strategic space priorities and presents a number of imperatives that are critical to enhancing and sustaining China's space power in the future.

To carry out a comprehensive analysis of China's space power capacities and its strategic priorities, it is important to situate its space programme in a national and historical context. China's early space efforts date back to 1956, when Mao Zedong, driven by the desire to 'rise beyond the imperialist legacy'[165] and restore international prestige and military capacities after a long period of war, launched China's nuclear weapons and space programme.[166] Planning for the Chinese manned spacecraft programme followed in 1966.[167] China also launched a variety of weather, communications and surveillance satellites in the 1970s.[168] By 2003, the country had conducted 67 successful satellite launches and become the third nation after the Soviet Union and the United States to send an astronaut into orbit around the Earth.[169] Although China's space history has been marked by many achievements, it has not been uninterrupted success. The Anti-Rightist campaign, the Great Leap Forward, the 1960 Cultural Revolution and the ideological rupture with the Soviet Union, which had strongly supported early Chinese space efforts, significantly undermined space research activities and reduced funding for science and technology. This undoubtedly retarded the growth of China's space programme and set it back many years, if not decades.[170] Yet, despite these traumatic political events, China's space programme continued to advance, albeit at a much slower pace than the Soviet and US programmes.[171] Deeply influenced by techno-nationalist ideas, Chinese officials considered technological progress, particularly in the field of space, a means to foster national economic growth, gain international prestige and reinforce the country's military capacities. Thus, China's space programme has survived difficult periods and even the shift in political paradigms from strict communism under Mao Zedong towards a more market-oriented communism.[172] According to Dr. Yanping Chen, the phase starting in 1986 marked the beginning of a prosperous era for China's space programme, with the government explicitly making space 'a cornerstone of the national science and technology development effort'.[173]

Table 5.3 China

Issue area	Space power capabilities	Space power imperatives
Social and health issues	China's civilian space applications derive from military assets and serve agriculture, transportation, telephony, television broadcasting services, tele-medicine and education.[152] Sustainable development and social progress are central to Chinese space policy.[153]	Given China's social and economic disparities and problematic demographics, a more balanced approach between civilian and military space priorities would better serve its social goals.
Domestic politics	The quest for domestic and international prestige underpins the whole Chinese space programme. Space exploration and manned space flight are considered highly rewarding activities and are therefore put at the top of the Chinese space agenda.[154] The White Paper highlights the 'cohesive force' of space 'for the unity of the Chinese people, in order to rejuvenate China'.[155]	With the risk of possible domestic unrest, due to uneven growth, political corruption and the struggle for resources in a context of problematic demographics, avoiding huge and expensive space missions and a focus on cost-effective and socially oriented space applications could better serve internal stability.
Economics	China's space programme serves the country's Five-year Plan for Economic and Social Development. China's space industry is seen as a central driver for economic development in the whole country.[156]	Improve the quality and reliability of commercial space applications and systems. Increase cooperation, both internationally and with the private sector, to leverage resources and reduce the risks of space operations.[157]

(*continued*)

Table 5.3 Continued

Issue Area	Space power capabilities	Space power imperatives
The environment	The construction of an Earth observation system, including meteorological satellites, resource satellites, oceanic satellites and disaster monitoring satellites, is part of the Chinese short-term development targets for the next ten years.[158] Given China's geographical and demographic particularities (mountainous relief, important deserts and a large population) observation satellite applications are critical for agriculture, natural resources and catastrophe management.	Support for international initiatives and increased information sharing will enhance China's capacities to address climate change issues, manage natural resources, mitigate space weather risks and improve responsiveness to natural catastrophes.
Science and human potential	China invests in scientific education and technical capabilities as part of its effort to improve the national economy. It is highly concerned about improving the worldwide reputation of Chinese technological achievements. The promotion of the high-technology sector and a well qualified space workforce are priority goals of the 2006 White Paper.[159]	Invest in research and development to improve its mastery of advanced space technologies. Encourage the young to embark on scientific careers to create a pool of highly educated experts.
Military and security issues	Confluence of China's military and civilian bodies and opacity of the space programme. China has used its asymmetric military space advantage (ASATs and ground-based laser),	It is in the best interests of China to refrain from conducting ASAT tests and from developing space-based weapons; emphasize multilateral rather than unilateral action and encourage

	particularly in its relationships with the United States.[160] It developed critical dual-use space capacities, with the GNSS Beidou/Compass expected to be operational regionally by 2013.[161] The international community perceives China's space activity as a threat to space security and suspects it of pursuing anti-missile purposes.[162] The protection of national interests in the military sense is a core function of the Chinese space programme.	confidence-building measures; support international initiatives to prevent the weaponization of space; and reduce the vulnerability of its space systems though enhanced recovery capacity and sharing of information.
International diplomacy	China plays an active role in the international space dialogue and advocates space arms control (PPTW), jointly with Russia. It is involved in many bilateral space cooperation agreements, namely, with the EU (Galileo), Russia, Brazil and Canada, and cooperates with emerging markets (Nigeria, Venezuela).[163] It is a strong regional space actor and has initiated the Asia-Pacific Space Cooperation Organisation (APSCO).[164] China's space programme is an important diplomatic tool for advocating its interests, including its relations with Taiwan, and increasing its influence on the international scene.	Emphasize international cooperation rather than unilateral action and encourage confidence-building measures. Take part in international space ventures such as the ISS.

During the past decades, China has stepped up its space activities and developed robust space capacities. It has built up important launch capabilities by converting its missiles into rockets and created three launch sites. In 2010, China launched a record 15 satellites and thus became the first country since the end of the Cold War to equal annual US launches.[174] Besides its launch capabilities, China has engaged in various satellite projects, including communications, remote sensing, and meteorological and science-oriented satellites. While all of them have been initially developed for civilian purposes, the information they collect and the technological know-how gained during their building process can obviously be transferred to military applications as well.[175] Hence, it is difficult to differentiate between Chinese civilian and military space activities, as most of the systems have dual-use capacities and some civilian programmes are placed under military supervision.[176] China has also developed its own navigation satellite system, Beidou, with a primary focus on military applications.[177] Given the strategic importance of such space technology, the Beidou project has received particular attention from China's policymakers and has been expanded to a more global programme, Compass.[178] In addition to satellites and launch capacities, China has achieved significant success in the space sciences and manned space flight, sending a man into space in 2003 and successfully repeating the experience in 2005.

China has a long-standing and expansive space programme that covers a large spectrum of space power capacities. As is noted above, in 2006, on the fiftieth anniversary of the Chinese space programme, China issued a White Paper in which it outlined its fundamental space objectives for the next five years. China's space programme is identified as a key strategic priority, and its budget is expected to grow substantially in the future.[179]

China's continuing quest for both domestic and international recognition and prestige has been a constant component of its endeavours in space and refers to the domestic politics capacity of *meta*-geopolitics. First of all, a successful space programme provides a great deal of legitimacy to the ruling Chinese Communist Party (CCP). As Kevin Pollpeter explains, 'by developing a robust space program and participating in high-profile activities such as human space flight, the Communist Party demonstrates that it is the best provider of material benefits to the Chinese people and the best organization to propel China to its rightful place in world affairs'.[180] Second, spectacular space accomplishments such as manned space flight can have significant geostrategic importance, as they yield international prestige and can be translated into stronger political power on the world stage.[181] Therefore, despite the important costs of human space flight Chinese officials continue to advance human space flight and space exploration as priority goals. China's future space ambitions include the launch of the Tiangong-1 module and its docking with the Shenzhou-8 spacecraft, probably in the second half of 2011, to form a space laboratory.[182] Chinese officials also talk about a robotic moon landing in 2012, as well as a manned mission to the Moon proposed for 2017.[183] Space docking is a complex technique necessary for the building of a space station and China does not hide its ambitions to create its own space station by 2020.[184] This latter goal is especially important to China, as so far the country has been prevented from participating in the development and missions of the ISS.

In recent years, China has clearly stated its willingness to join the ISS and it has advanced technical space capabilities as well as significant financial contribution to the project. Having the Chinese Shenzhou-8 spacecraft able to dock with the ISS would also reduce dependence on Russian spacecraft. China has hinted that it would like to be a part of the ISS project and that it would drop its plans for its own space station if it were given the chance to participate.[185] The Chinese manned space programme is a way for China to gain recognition from the international community. Such access would symbolize that China is a major space power, but allowing China to join the ISS would have significant geopolitical implications.

While China is investing heavily in space at the moment and would be a welcome contributor to the funding of the ISS, this could be more controversial than Russia's entry. There is strong ideological opposition to letting China enjoy the full benefits of such cooperation. Russia was initially brought into the project for counterproliferation purposes.[186] As a member of the ISS, Russia agreed to live up to the Missile Technology Control Regime (MTCR), and its aerospace engineers were provided with more secure employment with the ISS than in the chaotic Russian space agency of the early 1990s. There is a deepening of relations between Russia and China and a significant possibility that their own cooperation in space will expand. Pressure from Russia to allow China's participation in scientific experiments with other ISS partners, such as the EU, Japan and Canada, shows that it is difficult to exclude an emerging space power from a truly international cooperation programme. However, the international community, and particularly the United States, is very mistrustful of China and fears the risk of technology proliferation to countries such as North Korea and Iran. There is also a long-standing disagreement over the MTCR between China and the United States and these differences would need to be resolved before an international cooperation agreement on the ISS could become a reality. At the same time, Congress opposes US cooperation with China and despite all the pragmatic reasons (China's economy and developed space technologies), this would seem to 'put the brakes on any prospects for cooperation with China',[187] at least throughout 2011.

China's space programme benefits from a strong public relations component that helps improve its image abroad and at home.[188] China skillfully exploits media-friendly space events and it even qualifies as a 'useful propaganda tool' for the CCP.[189] Manned space flights and the 2007 lunar orbiter have been broadcast countrywide and the first space walk by a Chinese taikonaut was celebrated all across the country. Contests for schoolchildren to create artwork to commemorate the feat, and a rocket-shaped mobile device, painted red with *Shenzhou VII* stamped on the side are only some examples the nationwide pride in China's space achievements.[190] According to Johnson-Freese, '[a]ttention for a successful venture translates into prestige, with techno-nationalistic overtones. That is, it provides both a positive rallying event for the Chinese population as a whole, providing a sense of pride and achievement, with spillover externally in terms of technical achievements being equated to national power.'[191]

The success of highly complex space projects such as manned space flight obviously requires important technological know-how and a highly skilled workforce. The expansion of the high-technology sector and the creation of a pool of space experts are key priorities in the White Paper. Investment in a strong technical and scientific infrastructure has brought significant advances, some of which are specific to the space industry (re-entry lift control, emergency rescue, soft landing, heat prevention, system integration and testing), while others are more general (computer proficiency, electronic equipment, manufacturing technology) and therefore offer innovative technological spin-offs to non-space industries.[192] China's space programme educates and employs a whole new generation of engineers and scientists, thus significantly enhancing the human capital capacity of *meta*-geopolitics. Given the inherent dual-use nature of space technology, the availability of such a large pool of highly skilled experts serves both military and civilian space domains, while benefiting other sectors of the domestic economy and consolidating the world's recognition of Chinese technological excellence. A 2005 panel report commissioned by the US Congress estimates that China produces about 600,000 engineering graduates a year – as opposed to 70,000 in the United States.[193] Such estimates, together with statements of future Chinese space ambitions, add to US concerns regarding its technological superiority in space.[194]

China's space programme has historically been a primarily military endeavour, both in terms of funding and research goals.[195] In the 1950s, Chinese officials, thinking that prestige along with advanced military capacities 'would prevent a return to imperialist exploitation'[196] launched a large nuclear weapons programme. As a reaction to the use of nuclear weapons against Japan and US interventions in Korea and the Taiwan Straits crisis, China accelerated its nuclear weapon programme and long-range ballistic missile research, seen as a deterrence measure against the growing US threat.[197] The ideological split with the Soviet Union in the 1960s reinforced existing tensions and led China to develop its reconnaissance satellite technology to monitor activity on the Sino-Soviet border.[198] After the Gulf conflict, like other countries, China recognized the advantages space technology gave the United States and sought 'to develop space-based force enhancement capabilities as part of its military modernization efforts'.[199] Later, China's 2006 White Paper identified national security as a vital goal of the Chinese space programme, and space was recognized as a critical means of 'attaining comprehensive national military power'.[200] China's interest in and development of space-based military applications raised concern in the international community, in particular in the United States which closely follows Chinese developments in this field.

Nevertheless, it is important to stress that China's official position advocates the use of outer space for peaceful purposes only.[201] At the international level, jointly with Russia, it opposes the militarization of space and calls for a multilateral treaty and mechanisms to prevent space-based missile defence systems and a potential arms race in outer space.[202] Some have argued that the original attempts to stop the United States putting in place space-based missile defense systems were the result of Chinese worries that the 'US deployment of missile defenses, especially national missile defense (NMD), will negate its strategic nuclear deterrent, potentially forcing China into an expensive arms race that it cannot currently afford'.[203] China argues that a multilateral agreement to secure peaceful uses of outer space would avoid global space-based US dominance and impede the United States from targeting China with its sophisticated military capabilities. Nevertheless, the draft Prevention of the Placement of Weapons in Outer Space Treaty (PPWT) proposals do not regulate the development and testing of ground-based ASATs and allow their 'research, development, production and terrestrial storage'.[204] Consequently, a PPWT would not affect China's ASAT capacities, which were tested in 2007.

The modernization of China's military capacities focuses on the deployment of navigation and communication satellites. Its vast mountainous and desert areas mean that China needs reliable long-distance communication systems to command and control its troops.[205] However, due to the inherent dual-use nature of space technology, it is difficult to accurately estimate the extent of China's military assets in space.[206] Overall, Chinese space military technological capabilities are estimated to be behind those of the United States and other space-faring nations. However, the military operations of the latter are highly dependent on space assets and this dependence is perceived by China as a potential vulnerability of US power. China's military power is currently much less dependent on space assets than US military power. While this is a technical disadvantage, given the increased capabilities such space technologies can generate, it makes China less vulnerable to a potential space conflict. This could give China a 'potential relative near-term offensive advantage',[207] which means that China could attack more US space assets than the United States could attack Chinese ones. However, this relative advantage will diminish as Chinese space capabilities improve and its dependence on space technology increases. Therefore, instead of investing in expensive and complex military systems that would increase China's dependence on space and thus enhance its vulnerability and create the need for additional protective measures, China seeks to take advantage of the asymmetrical balance of military space power and exploit US vulnerability. Thus, China focuses on the development of independent technologies that would enable it to deny or disrupt US military access to and use of its space-based military systems.

The Chinese ASAT test, successfully carried out in 2007, demonstrated to the whole world that while China does not have the same military capacities as the United States, it is capable of challenging its supremacy by asymmetrical means. After China's ground-based ASAT test, it took a week for the Bush Administration to confirm that China had destroyed one of its own satellites, becoming the third country to achieve this objective.[208] China successfully destroyed its low Earth orbit weather satellite by launching a 'kinetic kill vehicle',[209] which, according to US intelligence and independent experts, was probably boosted by a two-stage medium-range ballistic missile. It is important to keep in mind that from a technical point of view, launching direct-ascent ASAT weapons is fairly simple. Such launch capacities alone are not as important as well developed guidance technology and precision manoeuvrability.[210] According to US intelligence, China might be lagging behind in terms of the precision guidance capacities of its space technology. Even so, this test was the first kinetic kill ASAT interception to take place in 20 years. In destroying its satellite, China created 2377 identifiable pieces of space debris, making it 'the largest debris-generating event on record'.[211] The Chinese Foreign Ministry confirmed that the test took place, but provided no explanation.[212] Off the record, some senior Chinese officials have acknowledged that China believes 'the weaponization of space [is] inevitable',[213] and that China must be prepared to respond to this development. Others believe that China's ASAT test was a warning against US space ambitions.[214] In the United States, some expressed their concern about China's peaceful exploration of space and argued that it is merely a means for building up China's military capacities, and, in particular, China's espionage capabilities.[215] In any case, the ASAT test generated considerable mistrust among the international community and turned out to be counterproductive for China's other important space goals, such as joining the ISS.

Besides national pride, and military and technological benefits, China's policymakers see in space a driver for national social and economic development. Although both official documents and Chinese officials repeatedly state that sustainable development and social progress are important objectives of Chinese space policy, little information on or evidence of such investment is available.[216] This is partly due to the secrecy of the Chinese programme, as the information on space costs and budgets is kept confidential.[217] Furthermore, as most space technology is dual use, Chinese civilian applications of space assets are often the result of an expansion from the military domain and therefore the distinction between the two is blurred. Space launch capacities, for instance, were developed from the foundations of Chinese IBMs. China's geographical and demographic particularities make it a major challenge for policymakers to 'feed its population of more than one billion on the arable land available'.[218] In this context, Earth observation satellites, apart from supporting military applications, are also widely used in agriculture, for monitoring the availability and distribution of natural resources and the prediction of natural disasters.[219] Communications satellites also serve numerous civilian applications and their added value has been highlighted by Chinese officials 'for education, government, transport and the financial and commercial sectors of the economy'.[220]

China's interest in commercial space activities emerged in parallel with its gradual move towards a new ideological pattern, dominated by the capitalist model. The shift took place in a difficult economic environment and was marked by a speech by Deng Xiao-Ping in 1978, in which he asserted that the national space programme must contribute to 'China's wider social and economic goals'.[221] Given this new economic orientation, China moved from initially dominant military objectives towards more commercially oriented space activities.[222] In the 1990s, it built up a successful space industry focused on developing and launching communications and surveillance satellites. An important role in the success of the Chinese space industry was played by its policy of the export of space-related products. China positioned itself as a 'low cost provider of space technologies and launch services' and used a package deal business model to export assets such as satellites, selling a package of satellites, operations and launch services.[223] Given that launch services are still rather expensive, this combination of products allowed it to decrease the overall price of the package through a reduction in the manufacturing and operational costs in the other steps of the process. As a result, China was able to offer packages that were maybe technically less sophisticated than Western technology, but also less expensive.[224] China, in turn, gained valuable experience in the sector, and could invest the gains in the expansion of other markets.[225] Taking advantage of its competitive prices and the temporary difficulties experienced by other providers, China launched satellites for numerous Western nations, including the United States.[226] Indeed, a lot of communication satellites that used the commercial launch services of China were built in the United States and, from that perspective, we can say the United States contributed to the development of the Chinese space launch industry.[227]

China's space activities may be less sophisticated and less well funded than US activities,[228] but they are catching up rapidly. Advances are often achieved by buying or copying technology, or through having a foreign partner transfer it 'as part of the price of access to a large potential market'.[229] Finally, it is important to point out that commerce in China is not solely a matter of the private sector, but is closely linked to the state's diplomatic agenda.[230] As Pollpeter points out, '[i]t is no coincidence that China's two satellite export agreements were signed with countries with large oil reserves – Nigeria and Venezuela'.[231] This point is discussed further below.

China's space programme is an important diplomatic tool for advocating its interests and increasing its influence in the international arena. By signing cooperation agreements with newly emerging space actors, such as Brazil, Venezuela and Nigeria, China has managed to increase its influence in the developing world, challenging the status and authority of the United States.[232] China created and hosts the headquarters of the Asia-Pacific Space Cooperation Organization (APSCO), which is a clear indication of China's aspirations for regional leadership in space affairs.[233] In addition to APSCO, China is engaged in many bilateral space cooperation agreements, mostly limited to civilian space activities. As opposed to its tense relationships with the United States, China enjoys good relations with Russia and cooperates with the ESA and Europe, most notably in the development of Galileo.

On a multilateral level, China, together with Russia, is the most vocal country in pushing for a complete ban on space weapons. China's commitment to the peaceful use of outer space has been noted in the CD on several occasions, and it was underscored in 2008 by the submission of the Russian-Chinese draft treaty on the Prevention on the Placement of Weapons in space. Yet, China is taking an ambiguous, two-pronged approach to arms control in space. It is strongly promoting an international treaty which would prohibit the deployment of space-based weapons but does not prohibit research, development, storage and testing of terrestrial-based ASATs such as those tested by China. Nor does it contain any provisions on dual-use space technologies. At the same time, China is continuing with its own space weapons R&D, working on kinetic energy ASAT technology, lasers and a number of other space-asset attack options – as a sort of 'insurance policy'. In other words, '[r]ather than directly confront the United States' space hegemony, China has sought to negate it through a policy of encouraging multipolar modifications to the international space regime',[234] without harming its own ability to threaten US space power.

While attentively monitoring China's developments in space, the Obama Administration is keen to engage in a constructive dialogue with Beijing on space exploration and security. As an indication of these cautious but cooperation-oriented dynamics, the United States and China issued a joint statement during the visit of President Obama to Beijing in 2009, announcing an exchange of visits between the two space agencies' heads in 2010 with the aim of promoting cooperation between China and the United States in space exploration, including manned space flight.[235] After his visit in October 2010, however, Charles Bolden stated that 'US-China space cooperation would proceed at a slow pace'.[236] In a more recent statement made during Hu's visit to Washington in January 2011, the issue of space was discussed again, and a Chinese delegation was invited to visit NASA in 2011.[237] However, despite these encouraging declarations, further US-Chinese cooperation in space has been put on hold by the US Congress, until at least the end of 2011.

The multifaceted analysis in this section shows that China, like the other states in this chapter, fully recognizes the numerous advantages space offers and seeks to strengthen its presence there in order to increase its influence in world geopolitics. China has already developed important space power attributes and has a long-standing, robust space programme. This is a source of national pride and a symbol of China's position on the international stage. Science and technology, including space R&D, are regarded as fundamental elements of national and world prestige, sustainable economic growth and citizens' well-being.[238] While China could become a potential competitor for the other space-faring nations, in particular the United States, there are still huge gaps between China and major space powers. Their space technology and capabilities are far ahead of China's in many areas and if China wishes to reduce this gap, enormous investment will be required. This investment might, in turn, draw resources away from other essential areas of technology and development.[239] As long as the Chinese economy continues to grow, it seems that China will be able to continue to increase its investment in the space industry and strive to position itself as an active and respected player in space.

While interactions between China and the rest of the world have been increasing in the past decade, its position towards the United States remains ambiguous and is watched with concern by a number of space experts. In order to ensure a safe and sustainable use of space and preserve global security it will be crucial to integrate China into international space activities such as the ISS, environmental monitoring and space sciences.[240] As Johnson-Freese points out,

> Encouraging China to participate in programs of our choice, in areas such as environmental monitoring and space science, is a better option than allowing China to focus on perceived threats to which it feels it must respond. [...] In a globalized world – and China is increasingly integrated to the rest of the world economically – countries which are connected with other countries will find it in their own best interests to maintain the system rather than perturb it. [...] While care must be given to how and how fast China is integrated in areas involving dual use technology, it can be done, and will ultimately increase the security of all.[241]

Yet, while there is a strong case to be made for increased cooperation with China, the secrecy of its space programme coupled with its 'single-minded zeal to forge a unique world-class military and space program'[242] ambiguous position on the weaponization of space have raised important questions in the international community. Unless China opts for more transparency in its space activities, there is little chance of new cooperation initiatives emerging in the near future.

Russia: Space power analysis

The Soviet Union was one of the world leaders in space until the early 1990s, when successive political and economic crises seriously undermined its space programme. The steady decline was reversed in the 2000s, when the country started benefiting from important gas and oil revenues. Supported by an improved economic situation and strong political will, space was put back on to Russia's strategic agenda, with Russia's intent on restoring its space power status on the international stage. Table 5.4 assesses the status of Russia's space capabilities and proposes a set of imperatives that would better serve the future enhancement and sustainability of Russian space power.

In the early phases of international space exploration, the Soviet Union seemed well positioned to win the global space race. It started the space race with its October 1957 launch of Sputnik-1, the first man-made satellite successfully placed in orbit. Weighing just 83.6 kg and only 58 cm in diameter, Sputnik took 98 minutes to orbit the Earth on its elliptical path. For such a small object, Sputnik had a huge geopolitical impact. The achievement of Sputnik captured the world's attention and caught the United States off guard. It also increased domestic pressure for the United States to invest more in its own space programme. These concerns were reinforced by the fact that the satellite the United States had been working on was not only slower in coming to fruition, but also less technologically sophisticated. Sputnik had a larger payload, and the US public and government feared that the Soviet ability to launch such satellites would translate into an ability to launch ballistic missiles carrying nuclear warheads from Europe to the United States.[269] Thus, Sputnik's launch almost single-handedly started the space race between the United States and the Soviet Union. This competition would ultimately become hugely emblematic of key Cold War tensions and rivalries.[270]

One month later, the Soviet Union successfully launched Sputnik-2. It had a larger payload than its predecessor, which was also the first instance of a living creature being sent into space. Although the dog, Laika, did not survive the mission, the Soviet Union used the event to gain even more attention and international prestige. The launch of Sputnik-2 was timed to coincide with the fortieth anniversary of the Russian Revolution. The programme went on to demonstrate that

Table 5.4 Russia

Issue area	Space power capabilities	Space power imperatives
Social and health issues	The 'enhancement of people's life quality' is an important goal of Russia's 2006–15 Federal Space programme.[243] It covers television broadcasting, mobile telecommunications, weather forecasting and emergency management.[244]	Increasing the budget and developing modern and effective socially oriented space services would help Russia meet the growing needs of its society.
Domestic politics	Space has always been a core element of Soviet prestige and internal cohesion. It is seen today as a critical means to restore Russia's former status as a space power and enhance national unity.[245]	Foster space-related scientific and exploratory research to restore the prestige of Russia's scientists and positively affect the 'nation's self-confidence'.[246]
Economics	Space technology is a core driver of the Russian economy.[247] Russia's commercial space sector is related to the government, benefiting from contracts and subsides.[248] Enhancing Russia's competitiveness in internal and global markets is a priority objective of the 2006–2015 Federal Space programme.[249]	Improve efficiency and organize the fragmented space industry sector into fewer, coordinated holdings.[250] Improve the quality of satellites and commercial space services.[251] Increase cooperation, both internationally and with the private sector, to leverage resources and reduce the risks of space operations.[252] Remove political and legal barriers to allow domestic commercialization of space-related services (communications, navigation and mapping).[253]
The environment	The lack of remote sensing orbital systems in Russia hampers 'nature management, hydrometeorology, and emergency monitoring tasks'.[254]	Maximize the utilization of micro-satellites for Earth sounding.[256]

	The building of a national Earth observation capacity is a priority goal of the government.[255]	Contribute to the COSPAS-SARSAT system.[257] Supporting international initiatives and increasing the amount of information sharing will enhance Russia's capacities to address climate change issues, manage natural resources, mitigate space weather risks and improve responsiveness to natural disasters.
Science and human potential	Russia's space sector faces a shortage of personnel. The crisis that followed the collapse of the Soviet Union eroded technical expertise and human capital.[258]	Rebuild solid technological and human potential to modernize Russia's space systems and enhance the competitiveness of its space services on the global space market.[259] Increased participation in international ventures would help Russia bridge the technological and human potential gap.
Military and security issues	Military space capability inherited from the Soviet era. Deterrent capability dependent on launch-on-warning and early warning satellites.[260] Despite a significant increase in the military space budget, funding shortages persist.[261] Russia sees unilateral development of anti-ballistic missile defences and the militarization of space as major threats to its security and national interests.[262] The 2006–2015 Federal Space programme makes the modernization and protection of its fleet of satellites (most of which are dual use) a priority, putting a special focus on the completion of the GLONASS system.[263]	It is in the best interests of Russia to refrain from conducting ASAT tests and from developing space-based weapons; emphasize multilateral rather than unilateral action; encourage confidence-building measures; support international initiatives to prevent the weaponization of space; reduce the vulnerability of its space systems through enhanced recovery capacity and enhanced cooperation and sharing of information.[264]

(continued)

Table 5.4 Continued

Issue area	Space power capabilities	Space power imperatives
International diplomacy	Russia is an important actor in the ISS. It is searching for strategic partnerships with the ESA, Ukraine, India and China. Russia, jointly with China, introduced a draft treaty prohibiting the deployment of weapons in space to the CD in 2008. Russia led an initiative at the General Assembly First Committee that resulted in the adoption of a resolution establishing a group of government experts to work on the advancement of TCBMs as of 2012.[265] Russia is highly committed to further strengthening and developing collaborative dynamics with its neighbouring countries and emerging space nations, looking to increase its international visibility.[266]	Further develop and exploit the Russian segment of the ISS with the United States and the other 19 states.[267] Cooperate on long-term international projects on flights to Mars and Venus.[268] Take the lead in proposing solutions to prevent the weaponization of outer space, thus improving Russia's world prestige.

long periods of weightlessness were survivable.[271] Timing major space missions with important political events became a signature of the Soviet Union. These back-to-back launches heightened Western anxieties and pushed the United States to dramatically accelerate its space programme. On 12 April 1961, the Soviet Union scored another major victory in the space race when it successfully launched the first man into space.[272] A Russian cosmonaut, Yuri Gagarin, spent 108 minutes in orbital flight in the Vostok-1 spacecraft.[273] Over the course of the Cold War, the Soviet Union managed to develop and deploy all categories of space-based systems, ranging from missile early warning and reconnaissance to communications and satellite navigation.[274]

However, after the Soviet Union's initial successes, the United States began to outpace it in terms of spending and achievements. Within a month of Gagarin's space voyage, the United States launched its first manned space mission, led by astronaut Alan Shepard.[275] This was a turning point in the space race, and slowly the Soviet Union lost its competitive space advantage, suffering a big loss when the United States successfully landed a man on the Moon in 1969. The relative decline of the Soviet Union's space programme was reinforced in the aftermath of its break-up when Russia was forced to divert more of its financial resources to social and economic development. Although Russia did advance its space programme in the years after the break-up of the Soviet Union, this was more due to the extensive R&D effort of the Soviet Union and the Soviet space legacy than any Russian innovation.[276] With Russia's main launch pad now located in Kazakhstan, Russia's launch rate collapsed, and launch vehicles and spacecraft had to continue operating at a reduced level.[277] Throughout the 1990s the key challenges faced by the Russian Government included safeguarding existing military programmes and preventing the decline of its space-related infrastructure.

In the early 2000s, however, Russia began to benefit from soaring world prices for oil and gas and renewed its focus on space.[278] By 2006, Russia's military budget had doubled compared to what it had been just six years earlier. A fair amount of that money made its way to the Russian space programme. In October 2005, the Russian Government adopted a ten-year Federal Space Programme, defining future Russian strategic objectives in space. This document allocated a budget of USD 10 million to space activities for the period 2006–2015 and is still in force today.[279] Although this ten-year budget was a notable increase for Russia, it was still less than the United States spent annually on its space activities. The Russian Government defines the overall goal of the programme as the satisfaction of increasing needs of space utilization, 'to achieve the objectives of the Russian Federation which cover economic, social, scientific, cultural and other areas, as well as for the benefits of Russian security'.[280] After a decade of decline, Russia launched an ambitious space programme covering all the seven capacities of *meta*-geopolitics, each of which is investigated in detail below.

The new Russian space programme places particular emphasis on the economic benefits of space. In Soviet times, priority was given to R&D, the advancement of science and reinforcement of the state's military capacities. The economic dimension of space, central to capitalist systems, was low on the list of the strategic objectives of Soviet officials. With the collapse of Soviet Union and the subsequent restructuring of the economy, these priorities shifted more towards income generation. In the late 1990s, space was seen as a means of stabilizing the Russian economy after the 'lost decade'. With the improvement of Russia's financial situation, the space industry was rediscovered as an economic and, beyond this, as a strategic goal. In the 2005 Space Plan, the Government defines the modernization of space infrastructure and the development and export of new, knowledge-intensive space technologies as core activities to increase the competitiveness of the Russian space industry and achieve 'the goals of doubling the state's gross domestic product within ten years'.[281] Yet, such an ambitious enterprise requires high levels of capital investment. In the absence of a functioning financial system, government political and financial support has proved indispensable.[282] According to a 2010 European Space Policy Institute (ESPI) report, over RUB 21 billion (USD 609 million) in public funds was injected into the Russian space industry in 2009–2010 despite the world financial crisis.[283]

The flagship project best supported by Russian officials is the completion of its satellite navigation constellation, GLONASS. The Soviet Union launched the first satellite in 1980 but the collapse of the Soviet Union meant that the nascent system suffered from a serious lack of investment as well as general mismanagement. However, determined to provide an alternative to the US GPS and take full advantage of the boom in the space services market, the programme has received important political and financial support from the Putin-Medvedev Administration. Although the system has not reached the technical level of the US GPS, recent improvements have made it operable, offering coverage over 90 per cent of Russia and 80 per cent of the globe.[284] Regardless of the financial and technical difficulties, Russia is determined to continue to operate and improve the system and is committed to increasing the programme budget.[285] At the same time, Russia is developing a new generation of satellites, GLONASS-K, intended to improve the accuracy and resilience of the system to the level of the US GPS or Galileo. Its deployment is planned for 2013 and the system ought to be fully operational by 2016. Such a new-generation navigation system could become a serious commercial competitor to the GPS and Galileo in the future global navigation services market.[286]

In addition to the above-mentioned economic benefits that a fully operational and reliable constellation would offer Russia, it would also have important strategic and political implications. Willing to assert Russia's influence on the international scene, the government has intensified its negotiations with neighbouring states and emerging space nations (Ukraine, India and China) to develop strategic partnerships for the improvement and supply of GLONASS, with the ultimate goal of directing their preference to Russian services instead of the US GPS or the EU's Galileo.[287] In sum, Russia seeks to utilize the growing importance of space to contemporary geopolitics in order to assert its regained influence and prestige. As de Montluc explains, 'power affirmation' through space is financially and ideologically supported by the Government in place, with the ultimate aim to restore internally the state and 'the unity of the Russian Federation' and externally 'the country's power and independence of action'.[288]

The renewed centrality of the Russian space programme has certainly fostered diplomatic and cooperative endeavours. Russia has a long history of international cooperation in space. While much of the Soviet Union's space programme – and specifically its ASAT programme – was provocative or competitively oriented, there were also examples of cooperation between the United States and its communist counterpart during the Cold War. For example, in 1975 a Soviet Soyuz and a US Apollo spacecraft docked and spent two days orbiting the Earth. During the joint mission, which capped several years of exchanging information and working together to overcome language and technical barriers, each crew visited the other nation's craft and performed scientific experiments, thus laying the foundation for the ISS.[289] This cooperative initiative was a huge confidence-building measure during a period of détente.[290] Although the Apollo-Soyuz space mission was intended to be the first of many, it was another 20 years before US and Russian astronauts collaborated in space again.[291]

Another signature of the Soviet and later Russian space programme was the Mir Space Station, the world's first consistently inhabited space research centre, which was occupied for nearly ten straight years. Mir provided a large scientific laboratory in space.[292] Research for Mir began in the 1970s, and Mir's first module was launched in 1986.[293] Importantly, the Mir Space Station was another key area in which the Soviet Union and the United States were able to cooperate and foster confidence, even in the face of bilateral tensions.[294] Over the course of the Shuttle-Mir programme, Russia's Mir combined its capabilities with US space shuttles, which provided transportation and supplies, and enlarged the living and working areas.[295] Mir's mission ended when it was brought back to Earth in 2001. Most of the station was destroyed on re-entry, but experience with Mir proved invaluable in planning and launching the ISS.[296]

After the break-up of the Soviet Union, Russia continued its cooperative endeavours in the space sector. In June 1992, the then President of the Russian Federation, Boris Yeltsin, and the then US President, George H. W. Bush, signed a bilateral space cooperation agreement with the aim of creating a closer relationship between the countries' respective national space agencies.[297] In December 1993, Russia joined the United States and other international players in the development, design and use of the ISS.[298] The ISS is among the largest and most complex scientific endeavours ever undertaken by humanity, and Russia plays a central role in this adventure.[299] The fulfilment of Russia's international commitments, including the funding of the ISS, is seen as an important indicator of Russia's reliability as an international partner and therefore is a high priority for the government. In its ten-year plan, it clearly states that a '[w]eakened presence of the Russian Federation in space will cause unavoidable violation of the international commitments, primarily with the CIS states, European countries, the USA, China, India and other states, thus providing negative effect on the international prestige of the Russian Federation'.[300] Therefore, to maintain a positive image of Russia in space matters, Russia plans to finish the assembly of its segment of the ISS in 2014, and by 2018 Moscow hopes to have introduced an advanced transport system and better engineering support for the space station.[301] One of the priorities for the Russian space programme is the replacement of the Soyuz space capsule, which transports the crew to the ISS.[302] It plans to double the production of Soyuz and Progress vehicles to support a six-person crew at the ISS and will transport US staff prior to the beginning of Orion missions.[303] Now that the United States has decommissioned and is replacing its space shuttle fleet, Russia's Soyuz will be the only solution for transporting crews and supplies to and from the station. This will offer Moscow powerful strategic as well as economic leverage, with the United States paying

USD 65 million per seat on the Russian shuttle.[304] While Russia's plans to continue updating Soyuz are ambitious, their fruition is somewhat distant due to a shortage of funding.

Russia remains highly interested in international cooperation in space, but seeks to keep control over the partnerships it selects and focuses on states that better suit its strategic priorities. Although the United States and Europe remain important partners, Russia is looking for new markets for its space assets and services, and is increasingly turning towards Asia.[305]

At the multilateral level, Russia has actively participated in the CD, where, as is noted above, it has proposed a treaty to prohibit the deployment of weapons in outer space and the use of force against satellites and spacecraft. This proposal was put forward with China, making these two states the main advocates of such a treaty. According to Russia's Foreign Minister, Sergei Lavrov, '[w]eapons deployment in space by one state will inevitably result in a chain reaction. This, in turn, is fraught with a new spiral in the arms race both in space and on Earth.'[306] He stated that a PPWT would 'eliminate existing lacunas in international space law, create conditions for further exploration and use of space, preserve costly space property, and strengthen general security and arms control'.[307] A new space treaty would require complex political and legal negotiations, and Russia is making positive first steps on this front, using these discussions to affirm its growing influence in the geopolitical space landscape. In addition to supporting international legally binding instruments, Russia is favourable to *the rules of the road* approach and has been advocating since 2005 a General Assembly resolution on transparency and confidence-building measures in outer space.[308] During the session of the General Assembly First Committee held in 2010, a Russian initiative, supported by China, the EU and the United States, resulted in the adoption of such a resolution.[309] The main provisions of the document encourage states to submit to the Secretary-General 'concrete proposals on international outer space transparency and confidence-building measures',[310] and requests the Secretary-General to present a compilation of all the proposals received at the next session of the General Assembly and to establish 'a group of governmental experts to conduct a study, commencing in 2012, on outer space transparency and confidence-building measures'.[311] The General Assembly approved the resolution at its sixty-fifth session,[312] and requested the Secretary-General to put in place the group of experts.[313]

Another area of increased attention for Russia is the socio-economic sphere of modern Russian society and its fast growing needs in terms of mobile communications, and information transmission through television and radio, covering its huge national territory including remote and sparsely populated areas.[314] During the Soviet period, the Soviet Union used to be the most active civil space actor, but the political and economic turmoil of the 1990s significantly reduced the number of its civil and commercial satellites.[315] This decline was reversed after Russia's economy recovery and the return of space to the national strategic agenda. The opening of the country to the world business market also profoundly changed the consumption and communication mode of the society and led to a boom in the high technology and communications sectors. The government fully recognizes that space technology, such as communications satellites, has enormous potential to help satisfy this constantly growing demand and thus improve the living conditions of the Russian population. The ten-year space plan states that further space development should focus on applications that will provide the population with modern and efficient communications technologies and a 'socio-oriented set of TV programs',[316] thus enhancing people's quality of life countrywide.

Another capacity of Russian space power closely related to social warfare and seriously affected by the lack of investment and performance following the dismantlement of the Soviet bloc is the environmental capacity of *meta*-geopolitics. In the ten-year plan, the Government highlights the importance of Earth remote sensing capabilities to monitoring environmental change, forecasting and managing natural disasters, exploration for natural resources, collecting geophysical and hydro-meteorology data and providing the population with accurate weather predictions. However these critical needs contrast with the existing limited Russian Earth remote sensing systems, which dramatically reduce the national capacity to tackle 'natural management, hydrometeorology and emergency monitoring' issues.[317] Moreover, the immensity of Russian territory, covering 20–30 million km^2, and the location of numerous strategic industrial objects, makes it vital for Russia to develop independent and reliable space technologies 'to support permanent ecological monitoring of the territory of the Russian Federation, as well as status control of high-priority objects'.[318]

The Government makes the building of such monitoring capacities a priority and emphasizes the development and replenishment of the country's orbital space constellations and the operational continuity of the Russian segment of the International Search and Rescue Satellite KOSPAS-SARSAT.[319] Benefiting from a level of high political support, translated into an increase in budgets, the most recent Ressours-DK satellite was made operational in 2010, while several types of optical reconnaissance satellite, which focus on cartography and detailed photography, are currently also operational in Russia. Despite these improvements, Russia still does not have uninterrupted coverage of the Earth from its satellites and projects are under way to bridge these gaps.[320] Satellite Earth observation is an application that can serve civil as well as military purposes, something that gives it even more weight among Russian strategic priorities.

Compared to the United States, Russia's conventional military operations are less reliant on space-based systems. It has a fleet of military satellites (early warning, imaging intelligence, communications, and navigation systems),[321] but the bulk of these were developed during the Cold War and reliability is not guaranteed for today's operations.[322] The modernization and development of Russia's space military infrastructure has been placed at the top of the national space agenda. In the case of Russia, its reliance on missile early warning satellites is critical to ensuring its deterrent capability and protecting its strategic forces. The development and placement of weapons in space, particularly those with anti-satellite capacities, is perceived by Russia as a serious threat to its overall military geostrategic posture.[323]

During the Cold War, the Soviet Union developed and deployed BMD and ASAT systems designed to attack satellites in low Earth orbit.[324] The development of the system began in 1963 and continued until the early 1980s, when the Soviet Union declared a unilateral moratorium on further ASAT tests. According to US experts, the Soviet ASAT system was neither flexible nor technologically advanced. Not surprisingly, US analysts pronounced the entire system inferior to the US system.[325] Although Soviet testing of ASAT weapons ended in the early 1980s, in a 1987 interview the Soviet Premier, Mikhail Gorbachev, admitted that the Soviet Union was conducting research into but had no intention of deploying an additional ASAT system.[326] Russia formally decommissioned its ASAT system in 1993.[327]

As for the other sectors, the Russian military space programme was severely underfunded during the 1990s. However, with the improved financial situation in the country the Government started to focus on modernizing and expanding Russian national military infrastructures.[328] This renewed interest was largely a response to US space policy, specifically with regard to the general context of Russian-US relations. Russian military and government analysts expressed serious concern regarding the potential military balance and the growing gap between the military capacity of Russia and that of the United States.[329] Most of the

US initiatives under the Bush Administration were perceived by Russia as a serious threat to its national security and fostered the development of the Russian military industrial capacity to undertake necessary response measures.[330] In early 2009 there were indications that Russia was re-evaluating its ASAT programme. According to Laura Grego, '[b]oth the US and the Soviet Union appeared to be hedging their bets by pursuing anti-satellite arms control talks while pursuing anti-satellite technology, albeit at a low level'.[331] The ASAT weapons testing moratorium that began in 1983 is still in place today, but space continues to gain importance for Russian military operations and national security capacities.

Core space military efforts, as set out in the 2006–2015 plan, focus on the development of dual-use systems, such as communications and navigation satellites. One of the central projects is to create domestic orbital groups to meet the demands of the country's safety and defence. Russia's authorities seek to develop space networks to cover the totality of Russian territory. Such a network would consist of space systems providing safe telecommunications services, navigation, coordination and time orientation, and expanding Russia's participation in the international community of those which provide similar systems and services.[332]

Another Russian preoccupation is to ensure independent access to and freedom of action in space, anytime it is needed. This entails developing a national network of launch sites and continuing with reforms of the rocket and space industries. In line with these objectives, Russia announced the construction of a new launch facility for heavy payloads in Vostochny and the development of Angora, 'its first entirely new post-soviet era rocket', expected to be completed by 2011.[333] The new spaceport, planned to be operational by 2015 and certified for manned spacecraft by 2018,[334] will decrease Russia's reliance on Kazakstan's Baikanur Cosmodrome.[335] In addition to the advantage of independent access to space, this spaceport is 'hoped to boost the Russian space launch industry's overall competitiveness. [...] the site is expected to become a space industry hub, generating more than 20,000 jobs on a long term basis'.[336] In the meantime, the policy places special emphasis on the maintenance and development of the Baikonur Cosmodrome facility, the control and operation of which were handed over in 2006 from Russia's military space forces to the civilian Federal Space Agency (Roscosmos).[337] It is the oldest space launch facility in the world, and has been a core component of Russian and Soviet space programmes since the launch of Sputnik-I.[338] Overall, since 2007, Russia has been active in such military space areas as communications, navigation, early warning, optical reconnaissance and signals intelligence.[339] Most of these were developed for military purposes, but all these technologies also have numerous civilian applications.

Russian space ambitions are huge and although they are benefiting from increased funds made available from the country's energy and armament sector,[340] they are seriously threatened by the lack of skilled human resources and technological know-how. During the Soviet period, science and technology were considered essential for the maintenance of the Soviet Union's superpower status, and institutional R&D received significant support from the state.[341] The highest priorities in the R&D sector were the support of military and space exploration, where technological sophistication and self-reliance were essential.[342] As a result, the Soviet Union produced a remarkable number of space engineers and technicians and acquired unique expertise in space affairs. The demise of the Soviet Union drastically altered the role of R&D in Russia. It faced significantly lower demand from industry and, subsequently, a reduction in budgets and manpower. Moreover, the lack of cash flow in the 1990s led Russia to open its national space sector to foreign customers interested in Soviet hardware and expertise, especially in the launch sector which had an excellent reputation and low prices.[343] Joint projects with Western partners were carried out in the 1990s with the aim of commercializing the former-Soviet launchers and intercontinental ballistic missiles. While these measures saved important branches of the national space industry, such as launches and manned space flight, the opening up of the Russian space sector eroded its technological potential and led to a leak of critical know-how to foreign countries.[344]

In addition to the technological challenge, Russia currently suffers from a critical lack of human resources. The reason for this is that the generation of experts of the Soviet space golden age retired and has not yet been replaced. As Charlotte Mathieu explains, '"the brain drain" – internal and external – has led to a lack of experts in their 40s to 50s who would be needed to train the next generation, and the younger generation are no longer attracted by space science and engineering careers'.[345] The 2006–2015 plan places important emphasis on the rebuilding of Russian technological and human capabilities, with 'technical retrofitting, the introduction of new knowledge-intensive technologies, quality enhancement and renewal of scientific and R&D personnel'[346] defined as initial steps to improving the human resources and technical capacity of its space power. This situation has led the country to search for new partners and cooperation opportunities to increase its expertise, acquire technical knowledge and rebuild a strong scientific sector. With regard to the latter, researching planets and other celestial bodies in the solar system is an important objective of the Russian space programme.[347] This includes searching for extraterrestrial life, studying the mechanisms of the Earth's evolution and finding applications for extraterrestrial resources.[348]

The comprehensive and multidimensional analysis carried out in this section shows that, historically, Russia was a leading space power, with long-standing and vast space programmes, ranging from military and civilian communications activities to satellite imagery systems and navigation to manned space flight. Russia owns multiple launchers and launch facilities that are able to deliver a variety of payloads to all orbits.[349] Taking into account its Soviet legacy and strong offensive strategic nuclear forces, Russia continues to play an important role in the geopolitical space environment. However, following the Soviet golden age, Russia's space sector suffered severely from a lack of funds and political commitment during the economic and political turmoil of the 1990s. With the improvement of the Russian financial situation and political stability, the Russian Government brought space back on to the national strategic agenda and has adopted an ambitious space programme for the next ten years, aimed at modernizing and rebuilding Russia's space power capacities. The increased focus on space has also been reflected in budgets, official documents, the stated intention to restore it to its status as a national programme, new policies for Cosmodrome launch sites and the reorganization of Roskosmos.[350] All these changes come in the context of the government wanting to reclaim ownership of the achievements of the Soviet era, with space seen as one of the means for integrating a post-modern society in Russia and increasing its influence in international geopolitics.[351]

Russia has also stepped up its international cooperation, particularly with regard to space transportation and scientific research. While maintaining good relationships with its classic partners, Moscow increasingly looks towards Asia and other emerging space nations. As a consequence, it has gradually intensified its relationships with China and India and has established cooperative dynamics with Brazil and South Africa. These countries seek to develop and expand their space activities and Russia provides them with technical support and joint projects.[352] With the recovery of its economy and an increased focus on space, Moscow tends to engage in 'multidirectional, balanced and pragmatic'[353] partnerships in space and this change in Russia's cooperation priorities will influence global geopolitical dynamics.

Although Russia has recently achieved a notable degree of success in space, numerous factors continue to hamper its ambitions. Despite an improved economy, the budget allocated to space remains low compared to the huge objectives set by the Government. Furthermore, the space sector still lacks modern technologies in key sectors and is in dire need of a supply of new, young researchers and experts. Moreover, political support for the space agency to implement its ambitious agenda is questionable, especially since the Government's long-term plans remain somewhat unclear. Finally, although Russia has increased its cooperative efforts in recent years, its commitment to true, lasting collaboration is questioned by some of its allies.[354]

Non-state actors: Space power analysis

Over the past few years, in addition to state actors, the commercial sector has emerged as a core stakeholder in space. Companies from all around the world are competing with leading US enterprises in a growing number of activities. In parallel, governments and space agencies are increasingly outsourcing their space services to private enterprises, while the huge revenues generated by the growth in civil applications are attracting more private firms to the space business. In other words, space has become a truly global enterprise.[355] Private space companies have built up extensive technical knowledge and expertise and are playing important roles in a wide range of space programmes, including environmental, health, scientific and military missions. They are capable of developing low-cost, complex space assets and systems in almost all areas including launch vehicles; telecommunications, such as radio, direct-to-home television (DTV), mobile telephony and remote sensing and navigation satellites. The assets and services they provide have almost universal coverage and contribute to empowering governments, organizations, space agencies and individuals all around the world. The commercial sector has thus enhanced its power in the contemporary geopolitical space landscape and '[a]ll evidence points to a continuation of this trend'.[356] In order to evaluate the impact of commercial space on future geopolitical dynamics, I have carried out a *meta*-geopolitical analysis of two leading space companies: The European Aeronautic Defence and Space Company (EADS)-Astrium and Boeing. The results of this analysis are summarized in Tables 5.5 and 5.6.

Case study: EADS-Astrium space power analysis

The European Aeronautic Defence and Space (EADS) Company is the world leading group in aerospace, defence and related services.

Table 5.5 Astrium

Issue area	Space power capabilities
Social and health issues	Astrium offers global communications coverage and Earth observation and navigation satellite technology for numerous civilian purposes. • For example, Spot Image provides nationwide or regional coverage, and colour and high resolution satellite pictures publicly available on line.[368] • Connect-D provides the full range of private telecommunication services worldwide, including landline phones, mobile phones and access to the Internet and WiFi.[369]
Domestic politics	Astrium's contribution to the emergence of joint ventures with universities, space agencies, NGOs and governments will encourage collaborative and inclusive dynamics within a community.
Economics	EADS and its subsidiaries represent an important share of the global space market. In 2010 EADS generated revenue of EUR 45.8 billion, with Astrium's contribution being EUR 5 billion.[370]
The environment	Astrium contributes to improving environmental intelligence for global leaders and citizens, through improved natural resources management, data mapping, agriculture and risk and disaster management.[371] • For example, Astrium and Infoterra are the main industry players in the EU GMES initiative. Involved at the systems architecture and instrumental levels of the first three Sentinel missions, Astrium also offers remote sensing data services and applications through Infoterra.[372] • Astrium is responsible for the development and manufacture of the ESA Earth Care explorer mission.[373] • Astrium is in charge of the supply of two Earth observation satellites and the establishment of the integration centre and training facility for the Kazakh ERSSS (remote sensing satellite system).[374] • Spot Image and Infoterra use the SPOT and Terra SAR-X satellites 'and a broad range of spaceborne and airborne acquisition capabilities' to offer a unique set of precise Earth imagery.[375]

Science and human potential	Through its extensive experience across all sectors of the space industry, Astrium contributes to the emergence of innovative and complex space technologies and the creation of a highly skilled space workforce. • In 2010 Astrium employed more than 15,000 workers worldwide.[376] • The German space agency (DLR) and Astrium jointly developed a unique radar satellite, TerraSAR-X. This technology provides Earth observation data of unparalleled quality and precision, particularly useful for scientific and commercial applications.[377] • Astrium is the prime ESA contractor for the Columbus space laboratory, one of the first European contributions to the ISS. Astrium coordinated 41 subcontractors from 14 countries, 'bringing together an exceptional combination of state-of-the art space engineering expertise'.[378] Columbus was delivered to NASA in May 2006 and was successfully docked with the ISS in February 2008. The laboratory allows the astronauts to carry out important scientific experiments.[379]
Military and security issues	Astrium provides secure communications, Earth observation and navigation satellite technology to the military worldwide. A few examples of its military applications and services are listed below. • Paradigm services is the global leading commercial provider of 'military band satellite communications and the only provider of end-to-end, hardened and protected satellite services, using X-band and UHF frequencies compliant to NATO standards'.[380] • MilSat is the main contractor for the German Armed Forces' satellite communications programme, SATCOMBw Step 2. In charge of the overall system design, integration and delivery of secure communications capacity, MilSat has been managing an anchor station and supplying commercial capacity to its German customer as of October 2007.[381] • Astrium GEO-Information Services has extensive expertise in geo-information and offers a multitude of services capable of helping the military address global security challenges.[382]
International diplomacy	Individually, Astrium does not yet play a role in international diplomacy. However, the commercial sector, collectively, is strong enough to influence international decisions in the field of space.

The group comprises four subsidiaries: Airbus, Astrium, Cassidian and Eurocopter.[357] For the purposes of the present work, I focus on the activities of EADS-Astrium (Astrium), third in the world and number one European supplier of civil and defence space systems and services.[358] Astrium covers three main areas of work and operates through three branches:

- Astrium is the global leader in the design and manufacture of satellites and related ground segment and equipment;
- Astirum Space transportation designs and produces space transportation for launchers, orbital infrastructure and manned space activities;
- Astrium Services develops and delivers satellite services.[359]

Internationally recognized for its technical expertise, Astrium leads a large number of flagship space programmes and delivers space services in many areas, ranging from social, environmental to scientific and defence missions. Astrium is prime contractor for the design, development and testing of the EU Galileo system and contributes to enhancing European economic, security, environmental and scientific capacities.[360] The company also played a central role in the development of Ariane launchers and is responsible for the coordination, manufacture and production of the new Ariane 5.[361] Astrium gained greater international visibility with the successful development and launch of the Automated Transfer Vehicle (ATV). Developed on behalf of the ESA, the ATV is an unmanned resupply spacecraft, designed to supply the ISS with propellant, water, air, payloads and experiments, and capable of reboosting the station into a higher orbit.[362] *Jules Verne*, the first ATV, was launched by an Ariane 5 and successfully docked with the ISS in April 2008. It was 'the first spacecraft in the world to carry out an automatic rendezvous and docking with a space station'.[363] The second ATV mission, *Johannes Kepler*, was launched in February 2011 and remained docked to the ISS for almost four months. It delivered equipment and fuel to the station and reboosted the ISS.[364] The success of *Johannes Kepler* opened the door for a third ATV launch, *Edoardo Amaldi*, planned for March 2012.[365]

Astrium's customers are governments, the military, commercial companies, space agencies and private corporations. Astrium's structure is complemented by its specialist subsidiaries: Surrey Satellite Technology (small satellites and sub-systems), Dutch Space (space instruments and simulation systems), Tesat (payload components), EADS Sodern (optronics and space equipment) and CILAS (laser technologies and high precision optics).[366] Astrium 'also provides launch services, through its shareholdings in Arianespace (Ariane launcher), Starsem (Soyuz launcher) and Eurockot (Rockot launcher), as well as space-based services via telecommunications and Earth observation satellites, through subsidiaries such as Paradigm, Infoterra and Spot Image'.[367]

Case study: Boeing space power analysis

Boeing is a global leading aerospace group and manufacturer of commercial jetliners as well as defence, space and security applications.[383] It operates worldwide through five divisions: Boeing Commercial Airplanes (BCA); Boeing Defense, Space & Security (BDS); Engineering, Operations & Technology; Boeing Capital Corporation and Boeing Shared Services Group.[384] For the purposes of the present analysis, I focus on the activities of BDS.[385] BDS provides services to the military, governments and commercial customers and covers a large range of space activities.

With its unique expertise, a turnover of USD 32 billion and 63,000 employees worldwide, BDS emerges as a central actor in the contemporary geopolitical space landscape.[386] It is the key contactor of the US Government and a central player in a number of critical national programmes, such as NASA's space shuttle and the ISS, the Missile Defense Agency's Ground-based Midcourse Defense program, and the US Army's Brigade Combat Team Modernization programme.[387] It is active in the economic, environmental, scientific and human potential, military and security global spheres with its space services and systems addressing 'the needs of warfighters on the battlefield, astronauts in space, meteorologists tracking a storm, or drivers navigating unfamiliar streets'.[388] Through its global coverage and extended expertise, the Boeing Company has acquired a set of soft and hard power attributes that are summarized and illustrated in Table 5.6.

Overall, the private space sector is emerging as a strong actor in the contemporary space landscape, with robust space power capacities that cover the seven dimensions of *meta*-geopolitics. It is an important provider of launch, imagery, communications and manufacturing services and its close relationships with governments and the military give it even more power on the international geopolitical stage. Increasing commercial competition drives technological progress and produces a highly skilled space workforce. Strong science and human potential capacities contribute, in turn, to the development of better performing and cheaper space products and services, more affordable to a larger number of consumers. The commercial space industry also has positive

Table 5.6 Boeing

Issue area	Space power capabilities
Social and health issues	BDS offers global communications coverage and Earth observation and navigation satellite technology for numerous civilian purposes. For example, BDS is building four 702MP satellites to refresh and expand Intelsat telecommunications capacity, 'distributing video, data and voice services from Asia and Africa to the Americas and Europe'.[389]
Domestic politics	Boeing's contribution to the emergence of joint ventures with universities, space agencies, NGOs and governments will encourage collaborative and inclusive dynamics within a community.
Economics	Boeing and its subsidiaries have an important share of the global space market. BDS has a turnover of USD 32 billion and employs 63,000 people worldwide.[390]
The environment	Boeing contributes to improving environmental intelligence for global leaders and citizens, through improved natural resources management, data mapping, agriculture, and risk and disaster management. For example, Boeing has recently delivered three Geostationary Operations Environmental Satellites (GOES) to improve the prediction and tracking of storms and other severe weather events, enabling earlier and more precise warnings.[391]
Science and human potential	Through its extensive experience across all sectors of the space industry, Boeing contributes to the emergence of innovative and complex space technologies and the creation of a highly skilled space workforce. • Boeing employs more than 159,000 people across the United States and in 70 other countries. It contributes to educate a diverse, talented and innovative space workforce.[392] • Boeing is the prime contractor for the design, development, construction and integration of the ISS. It manufactured the US portion of the station and is in charge of 'integrating the systems, procedures and components of 15 participating' states.[393]

	• Boeing is the developer and manufacturer of the US space shuttle, the world's only reusable launch vehicle capable of carrying out human space flight missions that allowed unique scientific experiments in space.[394]
Military and security issues	Boeing provides secure communications, Earth observation and navigation satellite technology to the military worldwide and serves global security purposes. • Contracted by the US Air Force Boeing is building '12 GPS IIF satellites for the US military's satellite-based radio navigation system to enable land, sea and airborne users instantaneously and precisely' to determine their location, speed and time 24 hours a day, worldwide and regardless of the weather conditions.[395] • Wideband Global SATCOM (WGS) provides enhanced communications capacity, coverage and operational flexibility for the military.[396]
International diplomacy	Individually, Boeing does not yet play a role in international diplomacy. However, the commercial sector, collectively, is strong enough to influence international decisions in the field of space.

externalities as the technology developed in the space sector is then often used in other industries, such as medicine, agriculture, and the manufacture of clothes or electronics, and enables the development of innovative and cheaper products in these branches as well. At the international level, the commercial sector makes space more accessible to all nations, including the developing countries, the populations of which, even in remote areas, have better access to information, better natural disaster awareness and management systems, as well as improved resource management models. Table 5.7 summarizes the space power capabilities of the private sector, treated here as an independent space actor. Like the state actors, the outcomes of the *meta*-geopolitics analysis serve as a basis for a set of imperatives that could increase and maintain the influence of private actors in space.

Satellite communications (Satcom) is one of the pioneers and nowadays core space business sectors. Initially developed by NASA, the establishment of the Communications Satellite Corporation (COMSAT) and the privatization of the International Telecommunications Satellite Organization (INTELSAT) transformed Satcom into lucrative commerce for telephony.[403] The telecommunications industry and the introduction of DTV have made Satcom one of the most profitable commercial space activities and created a vast market for satellite launch companies, operators and manufacturers.[404] As opposed to Earth observation or satellite navigation, the Satcom sector is almost purely commercial and now offers a large number of public service applications.

Whereas a few years ago, the use of space-based systems was the privilege of governments, satellites are now increasingly used by individuals, private firms and the world's media for data transmission, telephony, and television and radio broadcasts.[405] With the development of commercial Earth imagery, individuals with access to the Internet can easily surf on free platforms such as Google Earth, Google Maps or Yahoo maps to obtain high precision pictures of the globe, their country, or the area or house they live in.[406] The growing consumption needs of modern society have contributed to the emergence of a commercial satellite navigation industry. Extensively used in the aviation sector, navigation applications are now integrated into agriculture, construction, car navigation and mobile telephones. Initially devised for military needs, satellite navigation systems have become a core civilian and commercial space service, with European, Russian and Chinese systems seeking to challenge the supremacy of the US GPS in this strategic sector.[407] In the past decade, space infrastructure has rapidly

Table 5.7 The private sector

Issue area	Space power capabilities	Space power imperatives
Social and health issues	Space industry competition fosters innovation, lowers costs and makes space applications more efficient and affordable to a larger public. Private commercial space enterprises contribute to spreading the benefits from the exploration and use of space from rich to developing countries.[397]	Support private sector space initiatives that increase the capacities of the less-developed countries and the most vulnerable populations to deal with health, social, communications and educational issues. Further advance small space systems.
Domestic politics	The role of civil society in space is gaining importance and contributes to the development of advanced technologies, human potential and more cohesive and egalitarian societies.[398] Large space companies have gained important economic and military power and have the potential to influence domestic politics.	Engage in multi-stakeholder ventures, involving various community actors (commercial companies, universities, space agencies, governments, NGOs and individuals).
Economics	Space is a vast economic market, generating important revenues and offering high quality jobs worldwide. A global, competitive space industry stimulates technological progress and reduces the costs of access to space.[399]	Continued governmental support is needed to foster the demand for private companies' assets and services, especially in highly vulnerable sectors. A balance of legislation is essential to regulate the space business, while leaving enough freedom for the development of business-oriented initiatives.
The environment	Satellite monitoring of the Earth's environment provides a unique global data set and view of major climatic events and trends, critical for addressing environmental issues on Earth.	Provide services and data to help the international community address climate change issues, mitigate the risks of natural catastrophes and

(continued)

Table 5.7 Continued

Issue area	Space power capabilities	Space power imperatives
	The private sector has significantly improved the performance of the systems and the quality of data, while making the service less expensive and therefore more affordable.[400]	increase our understanding of the planet and its systems.
Science and human potential	The commercial space sector is a driver of technological innovation and scientific progress. It produces highly skilled engineers and managers. Scientific and industrial advances made in space have important spin-offs for other sectors such as medicine, pharmacology, electronics and metallurgy.[401]	Continued governmental support is needed to encourage private sector R&D. A balance of international legislation is essential to regulate the space business without hampering private sector R&D initiatives.
Military and security issues	Military capacities remain in the realm of state actors. Given the high level of interdependence between governments, the private sector and the military, the space industry indirectly supports military operations.	Refrain from activities that might contribute to the weaponization of space. Participate in space security multilateral mechanisms to advocate for the peaceful use of space and TCBMs.[402]
	To maintain the profitability of the sector, private actors have a direct interest in preserving space security.	Provide services and data for border control and conflict areas to help governments avoid conflicts and enhance global security.
International diplomacy	The private sector has become a key space player, with soft power capacities it can use to advance its interest in international space discussions.	Play a more active role in the international space debate. Provide services and data to improve treaty monitoring, and enhance confidence-building measures and global cooperation.

evolved to accommodate an embryonic private space flight industry. Russia has been the most prominent in this field, sending individuals to the ISS.[408] Commercially viable space travel has yet to become a reality, although many private investors continue to study ways to make this feasible.[409] For example, a company called Bigelow Aerospace is examining the possibility of creating a zero-gravity laboratory and even a hotel in space.[410] Access to new resources is also of vital importance today. Access to an infinite amount of minerals and raw materials and the possibility of endless clean and renewable solar power could dramatically change existing geopolitical dynamics between states. The economic potential and the possibilities for private commercial companies are tremendous in this sector.

In addition to the public consumption of civilian space applications, governments have played a central role in the emergence and further development of commercial space activities. The high costs of access to space, coupled with the potential of private companies to develop sophisticated and cheaper solutions, have reoriented governments' focus and investment towards the private sector. The space-faring nations have funded private R&D efforts and allowed the transfer of technology to selected private partners. Governments have subsidized private space programmes and advocated a regulatory regime that gives enough freedom of action to private actors to attract and maintain them in the space business.[411] As a result, the links between governments, business and space research have evolved into a complex of overlapping tasks, interests and relationships.

An important part of the commercial space industry's demand comes from government agencies, and it is critical for the future viability of the space business that government policies continue to stimulate this demand for private space-based technologies.[412] Space requires initial capital and entails uncertain long-term returns on investment. It also contains important risks and if the private sector does not have enough markets for its products, it will simply quit the business. Certain sectors, such as remote sensing or launch industries, are particularly dependent on government contracts, in contrast to satellite communications where individuals and private companies represent the bulk of the consumers and stimulate enough demand. The Obama Administration's decision to contract out human and cargo transportation to and from the ISS to private companies, following the retirement of the US space shuttle, is a good illustration of this hybrid private-public collaboration model.[413]

However, these private-public partnerships are not without risks, which are particularly high in sectors that rely heavily on government demand, and have limited alternative markets for their products. According to C. Homans, from *Washington Monthly*, the outsourcing of NASA's human space flight to the private sector typically entails these kinds of risks. NASA needs only few flights a year to bring astronauts and cargo to the ISS, which is not enough to cover the costs of a single launch company. The company will have to find other customers for its rockets, for example, in the space tourism or satellite sectors. The latter, however, is already very congested and the company would have to offer a truly innovative and cheaper service to enter this market, while space tourism is only at its embryonic stage. Homans concludes that 'if a company can't find other work, it has to spread its costs out over fewer launches to make its budget – which means the individual launches become more expensive'.[414] As a result, a solution supposed to be cheaper and allow savings could become even more costly. Homans also questions the capacity of a private launch company to launch people, which is quite different from launching cargo or satellites. He argues that 'while most rocket-building expertise resides in the private sector [...] most of the institutional knowledge of how to pull off a manned space flight safely remains inside NASA'.[415] He also raises the critical question of liability in case of a disaster.[416]

Governments are intensively partnering with private companies to achieve their military and national security objectives. This is particularly true for countries with military capacities that rely heavily on space-based systems. The US Government largely subsidizes the commercial remote sensing industry that helps satisfy its need for high resolution images, which are then diffused commercially with a lower resolution.[417] During the military operations in 2001 in Afghanistan, US forces used '700 megabytes per second of bandwidth, 75 per cent of which was from commercial systems'.[418] Hence, the commercial space industry is serving states' military and security objectives, which also leaves them more vulnerable to potential military attack and indirectly involves the private sector in the international global security debate. Given the important share of US industry in global space commerce, it would be particularly affected by any intentional or unintentional damage to its space infrastructure. If commercial assets are attacked or rendered inoperable by a collision with debris, their operators will incur significant financial losses. This is one of the strongest arguments for business and the private sector to play an active role in shaping global space policy and advocating against the weaponization of space. This business-military interdependence helps sustain US launchers and providers, but such strong links between the commercial space industry and government underline the need for the potential implications of the weaponization of space to be considered seriously by private sector actors.

Because space technology is mostly dual-use and its commercialization may disclose information vital for national security, governments agreed on export control measures to protect key national technologies and information. However, these measures can seriously hamper the development of commercial industry and affect its ability to participate in the global space market. Such restrictions are particularly marked in the US satellite industry. In 1999, the Government placed satellite export licensing on the State Department's US Munitions List, placing satellite product export licensing under the International Traffic in Arms Regulations (ITAR) regime and thus reducing the opportunities for US companies to participate in the world satellite launch and manufacturing business.[419] However, given the tremendous economic potential of private space activities, the commercial sector, in turn, is influencing US domestic politics, putting pressure on the Obama Administration and Congress to amend existing laws and regulations to make them more favourable to US commercial space activities.

The growing demand for civilian and military communication services, global positioning systems and remote sensing images is a core driver of the commercial space sector, which mostly focuses on building, launching and operating satellites.[420] These three sectors represent the bulk of the space industry today and are closely interrelated. Satellite manufacturers and operators need stable access to space, and the launch industry relies on the sustainability of demand from the satellite industry.[421] Today, Russia, the EU, the United States and China largely control the commercial manufacturing, operator and launch sectors.[422]

Since 2009, 'the satellite services industry has shown remarkable resilience to the adverse global financial conditions'.[423] The global nature of the business coupled with the growing demand for DTV, the Internet and mobile telephony in South East Asia and South America have allowed sustained growth. Moreover, the increasing outsourcing of governmental space services to private partners has resulted in the creation of a number of public-private partnerships, which have helped to protect the satellite services providers. Despite the emergence of new actors, the world satellite services business remains dominated by major US and

EU companies such as Eutelsat, Intelsat, Inmarsat, SES Americom-New Skies and Telesat.[424]

Among satellite manufacturing companies, US market share has dramatically declined since the introduction of the ITAR controls.[425] Currently, five companies dominate the global market: Boeing, Lockheed Martin Commercial Space Systems and Space Systems/Loral in the United States and Thales Alenia Space and EADS Astrium in Europe.[426] Other, smaller companies also play a growing role, such as Orbital Sciences Corporation in the United States; Khrunichev State Research and Production Space Center, NPO PM and Energiya in Russia; Mitsubishi Electric in Japan; and the Academy of Space Technology in China.[427] Small satellites are an important emerging technology in the manufacturing market and certain companies have focused on the development of this particular technology, which offers the same functionality as larger satellites but at a much lower price. This is the case for Instarsat, Microsat Systems, Space Dev and Swales Aerospace in the United States, Germany's OHB-System AG and the UK's Surrey Satellite Technology Ltd in Europe.[428] In parallel, manufacturers from China and India, with their pool of skilled, low-cost workers and the support of national space agencies and institutes, are emerging as providers of low-cost satellite technology.

Although launch capacity is vital to access space, it is more of 'an enabler rather than a significant economic activity',[429] and therefore generates less revenue than satellite manufacturing or operator services. The development of commercial access to space dates back to the 1980s, when the EU and Russia started using standard rocket technology, in contrast to the United States which was proposing to launch exclusively through its space shuttle.[430] Growing demand for launch services, coupled with a ban on commercial payloads on the space shuttle following the 1986 Challenger shuttle crash, encouraged competition in the commercial launch sector. Between 1988 and 1997 the French Ariane family of launchers covered about 50 per cent of the commercial launch market, joined by the Chinese Long March and the Russian Proton in the 1990s.[431] The US share of the commercial launch market decreased dramatically during this decade and has continued to decline in recent years. Competitors of the United States have proved equally reliable and less expensive, offering a better service to commercial clients.[432]

Currently, the commercial launch market is shared by three major actors. In 2010, Russian companies carried out 13 commercially contracted launches and generated revenue of USD 826 million, gaining 57 per cent of the global commercial launch market.[433] Russia's main companies are International Launch Services (ILS), which operates Proton and plans to use the new Angora rockets as soon as they are operational; the International Space Company Kosmotras (ISC Kosmotras), which operates Dnepr and Eurockot Launch Services, which uses Rockot.[434] Russia is closely followed by the EU, which has 26 per cent of the market, with the French company Arianespace and its Ariane-5 launch vehicle the only vehicle capable of carrying two satellites simultaneously. In 2010 it is reported that Ariane-5 placed 12 commercial payloads into orbit in 6 launches and generated revenue of about USD 1.32 billion.[435] The US providers have the remaining 17 per cent of the market with revenues of USD 307 million, mainly shared by Boeing Launch Services, Lockheed Martin Commercial Launch Services and Space Exploration Technologies.[436] In addition to the leading countries, other states, among which are Japan, China and India, have expressed an interest in entering the commercial launch market.[437] For example, the Antrix Corporation, the commercial wing of India's Space Research Organization, can provide launch services at a price that is roughly 70 per cent lower than the world market price, making India very attractive to foreign investors.[438] Increased competition among firms has contributed to reducing the cost of manufacturing and launching commercial satellites, thereby driving further investment and innovation in commercial satellite services.

The global space industry has seen steady growth over the past decade, a trend that is likely to continue for the foreseeable future due to growing commercial, governmental and military reliance on space-based systems and the need to reduce the cost of access to space. In 2010, while the global economy was slowly recovering from financial turmoil, the space economy flourished, generating an approximate total of USD 276.52 billion.[439] The bulk of this income was made by the commercial space sector, with a 13 per cent increase in the infrastructure and support industries (ground stations and equipment) to reach a total of USD 87.39 billion in 2010.[440] Satellite ground equipment has become a 'truly global market', with hundreds of companies from around the world producing components for the Earth stations and consumer electronics indispensable for the provision of satellite services.[441] Space products and services, stimulated by a growing demand for DTV, a global Internet, telephony and navigation systems, also makes up a substantial part of the space industry's business, reaching a total revenue of USD 102 billion in 2010.[442] This rapid growth in the ground equipment and satellite services sectors 'has offset the comparative declines in the satellite manufacturing and launch services markets'.[443]

The above-mentioned figures show the enormous economic potential of the fast-growing global space market. The commercial benefits are closely related to the state of technological know-how and the availability of a skilled workforce. Given the high costs and risks associated with space activities, private sector R&D relies heavily on government subsidies, but in return it allows important technological advances and produces a pool of highly skilled space experts. Developments in technology reduce the prices of space-related manufacturing, while cheaper electronic components and building materials provide higher profits for the companies investing in space businesses and create additional jobs.[444] Simply put, '"[m]ore space activity" translates into "more necessary infrastructure" and more "economic stimulus"'.[445] The benefits of the space sector then cascade to other industrial branches, thereby generating positive effects through numerous technological and economic spin-offs. In the end, everyone benefits from the development of a competitive and healthy commercial space sector.

The commercialization of space also has positive social consequences. Satcom technology for example, mostly commercial nowadays, can provide broadband services to remote areas and cover a large area, improving communications infrastructures. It saves public funds that would have been needed to increase the reach of terrestrial networks and offers a cheaper and better quality alternative for communications services. For example, Eutelsat launched a new generation of satellites, KA-SAT, in December 2010 capable of providing 'broadband access via satellite with the same level of quality as existing wired networks, for up to 1 million households out of the reach of terrestrial networks'.[446] Besides helping states address concrete societal issues, commercial industry offers a 'more egalitarian model than national space agencies' and gives developing countries a chance to participate in the fast-evolving and profitable space adventure. As Zach Meyer, from the Northwestern University School of Law, argues, space agencies are established to develop national space capacities and applications, whereas private companies offer their services to all customers, whether they belong to a space-faring nation or not.[447] The competition between private actors brings the cost of space down and improves the performance of space technology, enabling more countries, including in the developing world, to benefit from the exploration and use of outer space.[448] Thus, while the commercial sector is experiencing significant growth, space is becoming more globalized, with more countries entering space. For example, India is increasingly playing an active role in commercial launches and the manufacturing industry, offering lower cost assets and launch capacities. More developing countries such as Brazil, Nigeria and South Africa are focusing on the acquisition of satellite technology, including telecommunications and broadcasting services as well as navigation and Earth imagery, to encourage social and economic development in their respective communities.[449]

It is important to highlight here that civil society is taking an increasingly active part in the space sector, simultaneously affecting the domestic politics, science and human potential, and social capacities of *meta*-geopolitics. Globalization and competition between space actors have reduced the cost of space activities and enabled the development of smaller assets such as *smallsats* and *cubesats*.[450] These types of satellites are able to perform new tasks and can help further reduce production costs by their design and through large-scale production. They also require smaller and cheaper launch rockets and could be launched together with other spacecraft, with launch costs thus shared between multiple projects. These new and smaller satellite technologies have created incentives for space activities by non-traditional space actors and offered civil society a unique opportunity to get involved in space ventures. As reported in the Space Foundation 2011 Report, '[s]pace enthusiast communities are not mere observers, but are increasingly building their own in-space technologies through amateur or university satellite development programs.'[451] The R&D carried out by civil society, including private companies, not-for-profit organizations, institutes and foundations, has resulted in significant advances in science and the development of innovative space applications.[452] The increasing involvement of civil society in space 'not only stirs our imagination, but also brings us closer together – researchers, scientists, business professionals, and government officials',[453] thus gathering different actors of a community and encouraging collaborative work to advance common objectives in space. Such an active space civil society contributes to the emergence of a more inclusive and cohesive political and social order, fosters research and technological progress and eventually makes space more accessible to all nations.

Although the global space industry is growing, there is also an ongoing trend for mergers and the creation of strategic alliances. In the 1980s, there were approximately 20 major space companies in the United States. By the mid-1990s, the market was dominated by just three: Boeing,

Lockheed Martin and Northrop Grumman.[454] There has been a similar pattern in the EU. By the 1990s, two space conglomerates, EADS Astrium and Thales Alenia Space, dominated the market.[455] Russia's national space industry has also seen a trend towards consolidation. By 2006, several Russian space companies had merged into a new state-owned enterprise called Information Satellite Systems. The steady growth of the space economy attracts more stakeholders and puts pressure on private companies to offer competitive and cost-effective solutions, and mergers have become an important trend to cope with this competition.

The commercial sector is thus emerging as a space power, with certain diplomatic competences, capable of influencing world leaders in negotiations on space security and space international regulations,

> [C]ooperative efforts in this industry and the resulting coalitions that lead to cost-effectiveness in commercial space operations will likely be conducive to greater space access. If demand for space resources such as orbital slots and radio frequencies exceeds supply, as is starting to be the case, the result could be friction among providers of commercial services. However, such frictions need not necessarily be to the detriment of space security, as it could set the stage for a more coordinated and collaborative approach for the allocation of scarce space resources.[456]

In addition to the commercial sector, international NGOs are playing a growing role in the space security dialogue. They undertake advocacy, participate in international debates and organize events to share expertise and engage in constructive dialogue among international space actors. A number of them have the status of permanent observers at COPUOS and have an important role in linking with official actors and national delegations.[457]

Governments must continue to stimulate the commercial space sector, which is an important source of economic growth and employment, with positive spin-offs cascading to other branches of the economy. The decrease in the price of space activities and the development of smaller technologies opens the door to civil society, whose active role contributes to the emergence of more stable and cohesive communities and offers developing countries an opportunity to benefit from the utilization of space. The resulting competition for limited space resources has the potential to challenge the security and sustainability of space but could lead in the end to more confidence-building measures and better regulation in space.

Conclusions

The number of actors participating in space activities continues to grow and the enlargement of the space community raises new questions and issues. The fact that all the leading space-faring nations recently published new space policies shows that they fully recognize the importance of space to their national power and seek to reprioritize their activities and objectives to maintain and reinforce their influence in a fast-changing space reality. Examining the national policies and power capacities of space actors through the lens of *meta*-geopolitics helps to clarify the geopolitical implications of today's increased space activity. It allows certain predictions to be made on future developments and trends and helps devise tailored and sustainable recommendations.

Based on the *meta*-geopolitics analysis carried out throughout this chapter, the United States emerges as the leading player in space, with the technical and financial means to dominate the other space actors. The United States enjoys unequalled benefits from use of space to advance its national strategic objectives, with a clear priority given to national security and economics. As the leading space-faring nation, the United States has an important role in creating an environment where sustainable and responsible actions in space are promoted. However, US initiatives would be useless without the support of the other leading space powers, China, Russia and the European Union. Despite the softening in tone of its recently published national space policy, the United States remains unwilling to take any steps towards rules that might impinge its ability to act in space and it is similarly unwilling to make any commitments to refrain from using its capabilities in situations where its interests are threatened. Committed to maintaining its advantage in space, the heavy economic and military reliance of the United States on space-based systems makes it particularly vulnerable to asymmetric attacks, as is illustrated by China's ASAT test among other things. This dependence makes the United States, with its important commercial, military and civil sectors, the most vulnerable to both unintentional and intentional threats to such assets and it is in its own self-interest to keep space peaceful and free. This exposure partly explains the US focus on its military and economic space power capacities but also has the potential to compel the United States to engage in more cooperative dynamics on space security issues.

The US leadership position in space might be challenged by the other space actors, which perceive that US supremacy is a threat to their own security and development. The European countries collectively are emerging as a strong challenger to US commercial dominance. While they suffer from an inherent difficulty in coordinating a large number of national political views into a comprehensive joint policy, their level of cooperation is their main strength. However, the military and geopolitical importance of space technologies is complicated by European foreign and security policy, an area where progress has been slow and consensus difficult to reach. Despite this, the EU is taking steps to further its space power capabilities. The development of Galileo shows that it is attempting to break the US monopoly of the GNSS market and make itself less dependent on US technologies. As one of the most powerful actors in the space industry, the European model and its continuing encouragement of multilateral cooperation could contribute by enhancing the importance of space as a common sphere, enhancing cooperative behaviour and discouraging unilateral actions that could destabilize conditions in space.

Russia and China are also attempting to challenge US supremacy in space and strive to improve their relative positions as space powers. It is evident that after the break-up of the Soviet Union, the Russian challenge to US supremacy in space was no longer a serious threat. However, underfunded Russian space programmes have recently gained more support from Russian leaders and they are now expanding again. Its long history and tradition of space exploration has resulted in extensive domestic knowledge and experience and after the US Space Shuttle was retired in 2011, Russia now has a strategically important monopoly on sending people into space. In response to the US missile defence system, developed under the Bush Administration, Russia firmly opposed the weaponization of space and continues to enhance its international visibility by advocating a space weapons ban and the advancement of TCBMs. In parallel, Russia is committed to replacing its military space assets inherited from the Soviet era and is focusing on the development of dual-use applications, such as satellite navigation and Earth observation systems. It also cooperates with the EU, China, India and other space actors to strengthen its technological, human potential and economic capacities, which suffered severely from a lack of interest and investment during the post-Soviet period. Russia seeks to broaden its access to and use of space, thereby affirming its status as a major space power on the international stage.

On the same model, China made significant and impressive progress in such space technologies as satellites, launchers and manned spacecraft and has an ambitious plan for its future space programmes. Although it is officially committed to the peaceful and responsible use of outer space, its true intentions have been questioned since its controversial ASAT test in 2007. This test confirmed the Chinese determination to become a key player in space, but it also questioned its open commitment to the prevention of an arms race in outer space. China will not gain much by carrying out further ASAT tests. It has already demonstrated its capacity to destroy or disrupt satellites and such actions will create further suspicion from other countries, harm China's world image and delay its participation in international projects such as the ISS.

While the United States and Russia, together with the EU and China, remain the major space players, new countries and private sector actors are increasing their presence in space. Commercial space revenues have been steadily increasing over the past decade, with space becoming 'a global enterprise with the number of nations and firms with space goods and services growing rapidly'.[458] The private space sector offers services ranging from the manufacture and launch of satellites, to advanced navigation and communications systems. By reducing the financial barriers to space, the private sector has opened the doors of space to developing countries. The utilization of space by governments, businesses and civil society is growing, as our global community increasingly relies on space for communications, navigation, disaster management, environmental mapping, weather forecasting, military operations and monitoring climate change and scientific progress.

Overall, the space context, formerly driven by the competition between two superpowers, is evolving into a more complex and less predictable realm. Space power is gaining importance in world geopolitics, while the space landscape is witnessing the emergence of new actors, bringing space applications to almost every domain of our daily lives and raising new challenges, such as the interoperability of systems, orbit crowding, the weaponization of space and space debris. Outer space is often said to be a common heritage of mankind, without any national territory or the jurisdiction of individual states. The preamble to the OST recognizes the 'common interest of all mankind in the progress of the exploration and use of outer space for peaceful purposes'.[459] The common interest in space by all mankind implies a large degree of necessary cooperation, as it is not possible for anyone to avoid the negative consequences of hostile acts in outer space. Space is a common domain, and one irresponsible action will affect everyone. Just like the threat of mutually assured destruction from nuclear weapons, an extensive war in space would make space useless, and the decisions taken today will influence the use of space for many generations to come.

6
Space Governance and *Meta*-Geopolitics

Space opens up fascinating avenues and opportunities, but increased human activity in space entails new risks and presents important challenges for global security. Although the international community has repeatedly stated that cooperation and robust regulation are indispensable to ensuring the safety and sustainability of space, it also agrees that the existing legal architecture is not sufficient to address the new global challenges. In this chapter, I review existing institutions and legal instruments dealing with space governance. I also highlight the progress made since the beginning of the space age and identify gaps that need to be addressed. I also advocate two paradigms that are appropriate for space governance in our interdependent, globalized world and congested, contested and competitive outer space environment. I conclude by suggesting concrete measures that need to be taken by national policymakers, academics, international organizations, scientists, business, civil society and any other group engaged in the space sector in order to improve the safety of space activities and ensure the sustainability of outer space.

Existing cooperation and institutions

International institutions and cooperation efforts have increased in recent years, and space today is one international arena in which such institutions and efforts play a significantly greater role. Although the areas of cooperation were initially confined to the peaceful uses of space, such as astronomy, space exploration, commercial communications systems and meteorology, more recently, they have slowly expanded to cover more pressing, security issues such as orbital slot allocation and debris mitigation.

(1) The International Space Station

One of the most encouraging examples of international cooperation in outer space remains the ISS, funded by 11 European Space Agency (ESA) member states as well as the United States, Canada, Japan, Russia and Brazil.[1] The ISS has been a success in that it was accomplished as a peaceful cooperation between the former superpowers in what was one of the most strategic battlefields of the Cold War-space.

The concept of a permanent multilateral space station was developed at the initiative of NASA as early as 1981. The first phase of the plan was from 1981 to 1983, when NASA's Space Station Task Force convened space agency experts from the United States, Europe, Canada and Japan to discuss plans for a space station.[2] This was even before there was any great political will in favour of the concept. In the second phase, from 1984 to 1989, discussions were turned into genuine political commitments, which were formalized in a legal framework for what was then called the Space Station Program (SSP). The main reason for an SSP was to counter the Soviet Salyut and Mir stations. However, planning slowed at the end of the Cold War, most notably because of budget constraints. The resolution of the Cold War and the end of the space race put the plans in a new light. After the break-up of the Soviet Union, a third phase was initiated in 1993. This phase launched the collaboration that would be the ISS and included the integration of Russia into the project that also brought the benefits of Russian space technology and experience.[3]

Today, an international partnership of space agencies, led by the United States, Russia, Europe, Japan and Canada, produces and operates the components of the ISS, which is still under construction.[4] Segments of the station have been launched into space by the US Space Shuttle and the Russian Souyez launcher to be assembled in orbit. Only these two space launch vehicles were capable of taking cargo and crew to and from the ISS. After the last Atlantis mission in July 2011 and for the next few years, the Russian Soyuz craft will serve as the only way to reach the ISS. The ISS focuses on fundamental and applied research. The knowledge gained from the ISS is later applied in various domains, for example, to the development of new materials and new therapies in medicine.[5] Despite its numerous achievements, the ISS has been heavily debated and many still ask whether it is worth the cost. Indeed, the cost of the ISS is huge; the NASA development, assembly and running expenses (for the period of ten years) are estimated to be about USD 100 billion.[6] However, the ISS can play a vital role in promoting peaceful scientific and technological cooperation between critical players in space. Once it is fully completed, the ISS will contain 1200 cubic metres of pressurized space, which will be used for a large crew and a significant number of scientific experiments.[7]

(2) The United Nations General Assembly

The General Assembly is the main deliberative organ of the United Nations, 'where issues of international peace and security are collectively discussed among all UN member states'.[8] The resolutions of the General Assembly are not legally binding, but they are a demonstration of global opinion among governments and can have a normative value. They can indicate the establishment of customary law, international legal norms and guidelines for the appropriate behaviour of states. Some resolutions are adopted by consensus, demonstrating the clear agreement of the international community, and can lead to treaty-based negotiations. Resolutions that do not achieve consensus are adopted by a simple majority vote, apart from votes on issues considered central, such as peace and security and the election of Security Council members. Such issues require a two-thirds majority of member states, and issues dealing with space security are usually considered to be in this category.[9] Such votes show which states choose to remain outside a recommendation by making them reveal their position openly.

Within a month of Sputnik's launch, the United Nations General Assembly adopted its first resolution on space, setting out the non-legally binding principle of *the peaceful use of outer space*.[10] By December 1961, the General Assembly had adopted an additional resolution that establishes two leading principles for the exploration and use of outer space: (1) the application of general international law and in particular of the United Nations Charter to outer space and (2) the prohibition of national appropriation of parts of outer space and its resources.[11] Two years later, the General Assembly adopted the *Declaration of Legal Principles Governing the Activities of States in the Exploration and Uses of Outer Space* (the so-called *Principles Declaration*).[12] It determined the general legal basis for space activities. Once this regulatory foundation had been established, a number of space-related treaties, institutions and international conventions were slowly agreed on.

Since 1981, the First Committee of the General Assembly has passed an annual resolution calling all states to abstain from actions that threaten the peaceful use of space and supporting the negotiations in the Conference on Disarmament (CD) on a multilateral agreement for the prevention of an arms race in outer space (PAROS).[13] This is put forward each year with only minor technical and administrative changes. It is adopted by an overwhelming majority, usually with the United States as the only negative vote (or abstention) and Israel as an abstention. It shows the clear commitment of the international community to commence work on the issue of the weaponization of outer space. This is, however, a difficult task when the largest space-faring nation with the biggest capacity to develop space weapons consistently votes against such a measure. The move by the United States to abstain in 2009 might be interpreted as a positive sign that the new administration is more open to discuss PAROS in the future. In 2005, Russia launched a new initiative in the First Committee by introducing a resolution on transparency and confidence-building measures (TCBMs) in outer space activities.[14] The large majority of member states supports this resolution, with only the United States abstaining. It is important to remember that General Assembly resolutions are not international law. Only the Security Council can impose legislation on states. However, as is noted above, such resolutions are signals of general opinion among the member states of the United Nations and can contribute to the creation of customary international law.

(3) Committee on the Peaceful Uses of Outer Space

In 1959, the General Assembly set up the ad hoc Committee on the Peaceful Uses of Outer Space (COPUOS) to address issues related to 'scientific research, exploration, monitoring of the health of our planet, communications and navigation'.[15] COPUOS is one of the largest committees of the United Nations and is exclusively dedicated to advocating and promoting peaceful activities in space.[16] In addition to its 70 member states, a number of international organizations and 28 non-governmental organizations (NGOs) have observer status.[17] The decisions are taken by consensus and reported to the General Assembly, which usually endorses them.[18] COPUOS reviews the scope of international space laws and develops recommendations and programmes to encourage research and information sharing on outer space matters. It does not address issues concerning the military uses of space. Nor does it deal with the prohibition of the weaponization of space. However, considering the difficulties in distinguishing between military and civil applications, due to the development of dual-use technology, the decisions of COPUOS cannot avoid taking these issues into account.

It therefore maintains an open dialogue with the CD on the issue of the weaponization of space.[19]

COPUOS possesses two standing subcommittees: the Scientific and Technical Subcommittee (STSC) and the Legal Subcommittee. These two subcommittees form the basis for the substantive work of COPUOS, 'the only formal body empowered to negotiate new international space law'.[20] While the work carried out in these committees does not always generate substantive or quick results, some significant achievements have been realised.

The work of COPUOS has provided the international community with substantial contributions to the peaceful use of space, such as negotiating and producing the main international law on outer space. In addition to the 1967 *Treaty on Principles Governing the Activities of States in the Exploration and Use of Outer Space, Including the Moon and Other Celestial Bodies* (Outer Space Treaty, OST),[21] COPUOS has produced four other international treaties:

- The 1968 *Agreement on the Rescue of Astronauts, the Return of Astronauts and the Return of Objects Launched into Outer Space* (Rescue Agreement);[22]
- The 1972 *Convention on International Liability for Damage Caused by Space Objects* (Liability Convention);[23]
- The 1975 *Convention on Registration of Objects Launched in Outer Space* (Registration Convention)[24] and
- The 1979 *Agreement Governing the Activities of States on the Moon and Other Celestial Bodies* (Moon Agreement).[25]

Another significant accomplishment is the adoption of a set of voluntary guidelines on space debris mitigation in 2007. Building on the guidelines released by the Inter-Agency Space Debris Coordination (IADC) Committee, the STSC accepted a set of similar guidelines in June 2007,[26] endorsed by the General Assembly in January 2008.[27] The member states were asked to implement these guidelines through their national mechanisms 'to the greatest extent feasible'.[28] The guidelines consist of seven points, each with the intention of enhancing cooperation and decreasing the risks of debris in the future: (1) limit debris released during normal operations; (2) minimize the potential for break-ups during operational phases; (3) limit the probability of accidental collision in orbit; (4) avoid intentional destruction and other harmful activities; (5) minimize the potential for post-mission break-ups resulting from stored energy; (6) limit the long-term presence of spacecraft and launch vehicle orbital stages in the low Earth orbit region after the end of their mission and (7) limit the long-term interference of spacecraft and launch vehicle orbital stages with the geosynchronous Earth orbit region after the end of their mission.[29] This process took over five years, and the guidelines are only of a voluntary nature. However, they are a useful step forward, and the process provides a good example of how the international community can make progress towards a more comprehensive legal regime in the future.

Another step forward was the endorsement of a Working Paper by the Legal Subcommittee on the practice of states and international organizations in registering space objects.[30] The Working Paper 'recommends specific actions to improve state practice in registering space objects and adherence to the Registration Convention, including wider ratification of the Convention by states and international organizations, efforts to attain uniformity of information submitted to the UN registry, and efforts to address gaps caused by the ambiguity of the term "launching state"'.[31]

After its relative successes on debris mitigation and the implementation of registering space objects, the former Chairman of COPUOS, Gérard Brachet, argued for the expansion of the committee's scope.[32] Yet, revisiting the mandate of COPUOS to include all issues affecting the peaceful uses of outer space, including those concerning militarization, would not be an easy task. The United States in particular emphasizes a strong distinction between peaceful uses and non-armament, leading to potential complications in the negotiating process.[33] Despite this, Ambassador Brachet submitted a working paper to COPUOS with the objective of taking a deeper look at the critical issues that are challenging the future use of space.[34] The activities suggested include using space systems for a better understanding and monitoring of the Earth, coordination of global navigation systems, enhancing international cooperation in space exploration, supporting sustainable development with satellite technology, protecting or conserving designated areas of the Moon and other bodies in the solar system, further enhancing the work of the Action Team on Near Earth Objects, consideration of the non-technical aspects of future commercial space transportation and developing rules of the road for enhancing the long-term sustainability of space activities.[35] This proposal was followed by calls from representatives of the European Union for a more specific code of conduct in space.[36]

After two years of discussions, in June 2009, at the fifty-second session of COPUOS, the Committee agreed to add the concept of '*long-term sustainability of outer space activities*' to its agenda for 2010, under which COPUOS discussed ways of ensuring safe and sustainable access to and uses of outer space.[37] In February 2010, the STSC set up a working group with a mandate to develop best practices for safe and sustainable use of space. These *rules of the road* will encompass the contributions of member states, NGOs and the commercial space sector.[38]

The last treaty to be negotiated in COPUOS was the Moon Agreement. Since then, COPUOS has only been able to negotiate soft law arrangements, such as the debris guidelines and other non-binding documents. By pursuing careful soft law regulation, the international community can move slowly towards building a regime of space rules. This process is slow and needs to contain some confidence-building measures in order to reduce scepticism and mistrust, but the steps forward by COPUOS show that such voluntary commitments are easier to agree on than fully blown treaties. Scientific and commercial cooperation in space continue through the ISS and the International Telecommunication Union (ITU), but there is an urgent need for enhanced cooperation at a higher political level. Space policies must adequately address the threats we face today, and they must address these threats through a common perspective in which multilateral efforts are needed to secure each individual state's assets. COPUOS is one of the most appropriate forums in which to discuss these issues, and one where the international community can agree about steps to enhance and secure a sustainable way of enjoying the benefits of space.

(4) The United Nations Office of Outer Space Affairs

The United Nations Office of Outer Space Affairs (OOSA) is based in Vienna and is the central international organization dealing with the regulation and administration of international space affairs. It focuses on promoting international cooperation in the peaceful uses of outer space and also performs more technical functions, such as maintaining a register of the objects and satellites sent into space by different members of the international community. The organization initially served as the secretariat for COPUOS. It is now responsible for implementing the United Nations Secretary-General's responsibilities under international space law, maintaining the Register of Objects Launched into Outer Space under the United Nations space applications programme and implementing COPUOS decisions and recommendations.[39] The OOSA has two sections: the Committee Services and Research Section and the Space Applications Section.

The Committee Services and Research Section focuses on serving as a secretariat for the different committee meetings and conferences. It also provides reports and studies for COPUOS and its subsidiary bodies. These range from background information to extensive studies in fields such as space research, the applications of space technology, space law and international cooperation in these fields. In addition, the section also manages some tools for member states, such as searchable online indexes for the Register of Objects Launched into Outer Space and for the status of United Nations Treaties governing activities in outer space.

The OOSA organizes and carries out the United Nations Programme on Space Applications (PSA). This Programme was initiated in 1971 to 'create awareness among policymakers and government agencies of the benefits of space technology and to assist people from developing countries in acquiring the knowledge, skills and practical experience necessary for their application'.[40] In 1982, the programme was expanded and its mission was broadened to 'enhance the understanding and subsequent use of space technology for peaceful purposes in general, and for national development, in particular, in response to expressed needs in different geographic regions of the world'.[41] The end of the Cold War and the increasing number of actors in space made the need for enhanced cooperation even more important. The PSA focuses its efforts in key areas, such as space science, the utilization of global navigation satellite systems, natural resources management and environmental monitoring, satellite communications, space technology and disaster management.[42]

Some recent achievements include the establishment of regional centres for space science and technology education in India, Morocco, Nigeria, Mexico, Brazil and Jordan. The PSA has also facilitated the establishment of telescope facilities in Colombia, Egypt, Honduras, Jordan, Paraguay, the Philippines, Sri Lanka and Uruguay. However, organizing seminars, workshops, training courses and meetings is its main area of activity, and these are meant to increase understanding and experience of space applications, with the specific aim of reducing the gap between the industrialized and developing countries. The OOSA also maintains close cooperation with other space-related organizations, both international and regional, such as the Committee on Space Research (COSPAR), the International Astronautical Federation (IAF) and the International Institute of Space Law (IISL).

The United Nations and its institutions have used space applications increasingly in recent years and such applications make important contributions to their work. However, the United Nations does not have an independent space policy. Suggestions have been made that the OOSA should establish an ad hoc monitoring body, which would monitor debris and other objects in orbit.[43] This would be carried out through extensive access to data and catalogues of debris from its member states. Such a body could also verify and monitor compliance with such international space agreements as the OST, the Registration Agreement, the Liability Convention and the Space Debris Mitigation Guidelines. A more active United Nations could facilitate and encourage a sense of common ownership of space, due to its role as the principal universal instrument of multilateralism. The United Nations is one of the few organizations that can drive international and regional cooperation while also promoting an updated regulatory framework for such cooperation in the interests of all of mankind.

(5) The Conference on Disarmament

The CD was established as a multilateral negotiating forum in 1979.[44] It is not a formal United Nations agency, but it is closely linked to the UN. Currently it has 65 members and its mandate is to negotiate disarmament and non-proliferation treaties. One of its key areas of expertise is promoting the peaceful use of outer space and, since the CD deals with arms control and disarmament, it is an appropriate body for discussing weapons and the militarization of space. In 1982, the Mongolian People's Republic proposed to create a committee to negotiate a treaty that would fill the gaps of the OST as regards the weaponization of space.[45] PAROS was inserted as an agenda item in 1982 and three years later an ad hoc committee on PAROS was put in place with a mandate to examine the issue of an arms race in outer space.[46] Dissolved in 1994, the discussion of PAROS continued informally within the CD.[47] However, '[p]roblems related to the working method to be followed and issues to be addressed led the activities of the *ad hoc* committee as well as efforts undertaken by the CD after its dissolution, to failure and complete standstill'.[48]

In fact, there are several complications that hinder the work of the CD. One of the most serious is procedural. According to CD rules, the institution operates by consensus. This means that agreements cannot be reached if a single state objects. While the CD has negotiated important treaties in the past, such as the Non-Proliferation Treaty (NPT), the Biological and Toxic Weapons Convention (CBTW), the Chemical Weapons Convention (CWC) and the Comprehensive Test-Ban Treaty

(CTBT), it has not been able to commence work on any specific issues more recently. The conclusion of the CTBT in 1996 was the last accomplishment of the CD and it has since suffered over a decade of deadlock, resulting in doubts being raised about its effectiveness. However, procedural complications are often a sign of more substantial challenges. Since different states have different priorities, the deadlock has been caused by disagreements over what to work on first. While China and Russia prioritized negotiations on outer space, the United States only considered a fissile material cut-off treaty to be ripe for negotiation. At the same time, countries from the Non-Aligned Movement are vocal concerning negative security assurances and nuclear disarmament.

After years of informal negotiation and some welcome shifts in government positions, the CD broke its deadlock on 29 May 2009 by adopting a programme of work. The CD decided by consensus that it would establish a working group to 'negotiate a treaty banning the production of fissile material for nuclear weapons or other nuclear explosive devices'.[49] Working groups were also established on issues related to the prevention of an arms race in outer space, negative security assurances and nuclear disarmament, but without a mandate for negotiations. These three working groups were instead asked to 'discuss substantively, without limitation'[50] the issues concerned. This showed a clear change in position by the United States, Russia and China. A few years earlier, the Russian and Chinese governments had insisted on linking the negotiation of a fissile material cut-off treaty to discussion of a space weapons ban. The United States in turn refused to even discuss the issue of weapons and an arms race in space since, according to its delegation to the General Assembly First Committee, '[o]ne: there is no arms race in space. Two: there is no prospect of an arms race in space. Three: the United States will continue to protect its access to and use of space.'[51] The United States did not change its position on this issue until the Obama Administration came to power, and accepted the discussion mandate in the CD. The language in the established programme of work was delicately crafted, and on the issue of the weaponization of outer space it was a compromise between the states that want to start negotiations on a formal treaty immediately and those states that do not think that such formal negotiations are an attainable goal at the moment.

However, this fragile consensus did not last. Almost immediately after it was reached, Pakistan withdrew its support for the agreed programme of work. This decision was related more to another sensitive issue in the CD, the proposed ban on the production of fissile material for nuclear weapons purposes, than issues directly related to outer space. However, due to the procedural structure of the CD, formal discussions on outer space cannot commence until procedural details concerning negotiations on fissile material for weapons purposes have been resolved. Recently, another problem further exacerbated the frustration with the lack of progress in the CD, seriously undermining the credibility of the institution. Following the CD tradition of alphabetical rotation of its chairmanship, the Democratic People's Republic of Korea took the presidency in June 2011. While the United States 'has effectively made light of the matter, Canada has strongly registered its opposition to the very principle of North Korea holding the presidency, and is boycotting the CD'[52] for the succeeding two months. As a result, the CD remains deadlocked over its proposed agenda of work with no agreement in sight in the near future.[53]

Despite its procedural problems and even though the CD has been deadlocked for many decades, it remains the central global disarmament negotiating forum. It was in the CD that Russia and China presented their draft Prevention of the Placement of Weapons in Outer Space Treaty (PPWT) on 12 February 2008. The text is based on an earlier proposal introduced to the CD by Russia and China in 2002, basically stating that the existing space law regime is unable to prevent an arms race in outer space and making a case for a new treaty. The PPWT prohibits the deployments of weapons in outer space and provides a definition of space weapons.[54] However, this joint proposal has been criticized for its lack of provisions on ground-based ASAT tests. Additionally, issues persist with the utilization of dual-use space technology, which is not covered by the PPWT definition of space weapons. Another critical weakness of the draft is the absence of reliable verification measures, which significantly reduces the capacity of the treaty to ensure compliance with its norms.[55] The United States dismissed the proposal in an analysis of the draft, arguing that it was 'a diplomatic ploy' by the two nations to gain a military advantage.[56] Because the CD operates by consensus it has not been able to act on the draft treaty. In August 2009, China and Russia jointly submitted their working paper addressing the issues raised and comments made by the United States and the other CD members. Although the Obama Administration seems to be more open to discuss the weaponization of space, given the lack of verification mechanisms in the PPWT and the inherent difficulty of verifying such a treaty, it is likely that the United States will not pursue it. Furthermore, there is little chance that the United States will support a treaty that does not include terrestrial-based ASATs while China, Russia, India and other countries are increasingly developing such capabilities.

Russia and China have recently announced that they might consider a provision banning ASATs as possible amendment to the PPWT, a measure that would increase the chances of success of the treaty.[57] The security situation in space is becoming more serious and threats against the continued use of space are increasing. Discussions in the CD can make a significant contribution to a multilateral dialogue, which will be of assistance to space-faring nations and their future activities.

(6) Other multilateral space agencies and cooperation initiatives

The European Union and the United States also cooperate in the framework of the Columbus laboratory module and the Automated Transfer Vehicle.[58] Other examples of cooperation include a NASA agreement with India on technology exchange, an agreement between Russia and India to jointly use Russia's navigation system and, finally, an agreement between the US National Oceanic and Atmospheric Administration (NOAA) and the European Organization for the Exploitation of Meteorological Satellites (EUMETSTAT) to share meteorological information in case of war.[59] Almost all space agencies have cooperation agreements with other agencies, but the scope of these agreements is not always substantial.

Another important milestone in global cooperation in space is the Group on Earth Observation, which provides the Global Earth Observation System in which countries and international organizations carry out environmental monitoring and natural disaster warning and work to improve countries' economies.[60] It was launched at the 2002 World Summit on Sustainable Development by the G8 countries and is a voluntary partnership of governments and international organizations. By July 2010, the Global Earth Observation System had 86 member states plus the European Commission and 61 intergovernmental, international and regional organizations as participants.[61]

In December 2005, the General Assembly established the International Committee on Global Navigation Satellite Systems (ICG) with the aim of promoting international cooperation on issues 'related to civil satellite-based positioning, navigation, timing, and value-added services, as well as compatibility and interoperability among the GNSS systems'.[62] The ICG is an informal body, created by a General Assembly resolution in order to encourage cooperation on matters of mutual interest related to civilian satellites and global navigation satellite systems for sustainable development. The members of this body are states and organizations that are either GNSS providers or users of GNSS services. The ICG organizes the Providers Forum, a meeting where countries that operate GNSS systems or are planning to develop such systems can convene and discuss coordination and cooperation issues, such as compatibility, information dissemination and enhancement of performance. This meeting is an opportunity for countries with operational or planned systems – the United States (GPS), the EU (Galileo), Russia (GLONASS), China (Compass), India (Gagan/IRNSS), Japan (MTSAT), Nigeria (Nig ComSat-1) and the Asia-Oceanic regional initiative (Quasi-Zenith Satellite System) – to discuss issues such as how to protect the GNSS spectrum and mitigate the risks related to orbital debris.[63]

The United Nations Platform for Space-Based Information for Disaster Management and Emergency Response (UN-SPIDER) was established by the General Assembly on 14 December 2006. Designed as a central hub for satellite imagery requests during disasters, it aims 'to provide universal access to all countries and all relevant international and regional organizations to all types of space-based information and services relevant to disaster management'.[64] Through UN-SPIDER, countries can request and access space-based information that will complement the full cycle of disaster management from risk assessment to preparedness to mitigation. This programme especially benefits developing countries that are vulnerable to natural disasters such as earthquakes or cyclones but do not have their own space capabilities for predicting and tracking such phenomena.[65] UN-SPIDER is being implemented by OOSA and the German Aerospace Centre (DLR).[66] Other related international initiatives use space-based information for immediate emergency response, but UN-SPIDER is unique in its ability to offer support to all stages of the disaster management process, such as planning and risk reduction.[67] UN-SPIDER is involved in 'the mediation of worldwide access to existing sources of space information and services',[68] through fostering alliances and disseminating knowledge.[69] A web-based knowledge portal, international workshops, expert meetings and advisory missions are all used to achieve UN-SPIDER's goals.[70] Given that the world's population is expected to reach 9.22 billion by 2075,[71] it is clear that there will be a growing need for sustainable management of natural resources and that this will present numerous challenges relating to food and water shortages, the need for energy supplies, health care, and urban and rural transport as well as the need for climate protection, and security and disaster management. UN-SPIDER will play an important role in this respect. Although the use of satellite-based applications for social purposes is at its early stages, the applications have the potential to offer numerous benefits in the short to medium term.

There is a large amount of cooperation in space and many international institutions deal with the issue, a clear sign of its special characteristics. Such cooperation increases the incentives for preserving a common peaceful environment in space, and also enhances the abilities of developing countries to explore and enjoy its benefits. However, there are clearly some significant difficulties with such cooperation. The dual-use dilemmas of space technologies, together with the extensive military and security-related capabilities they entail, make such technologies a crucial part of national security. As a core of a state's sovereignty, national security issues are inherently difficult to submit to international cooperation agreements and institutions. Although the Cold War with its ideological conflict between two major space powers is over, security challenges remain for all countries. Space technologies might be even more appropriate to deal with the new types of threat we face in the twenty-first century and they are considered to be of the highest importance to current military operations. As with other agreements on arms control and limitations on military tools, international space cooperation is surrounded by controversy and suspicion.

The current lack of progress may seem disappointing. COPUOS was successful in creating international binding law to regulate space activities in the 1960s, but since the conclusion of the Moon Agreement in 1979 no further treaties have been negotiated. The CD remains deadlocked. However, the spin-off effects of successful cooperation in other space endeavours could make a contribution to progress in these areas as well. COPUOS has adopted a set of debris mitigation guidelines and there are ongoing discussions and pressure on COPUOS to continue work on other soft law rules such as guidelines, codes of conduct, and transparency and confidence-building measures to further increase the safety of space operations. The proposals for creating *rules of the road* and directions for safe and sustainable space activities could further increase the soft law commitments of states to comply with international rules.

Gaps in existing international space law

International space law has been developing since the start of the space age, and many of the above-mentioned institutions have had a major role in these developments. COPUOS has played a central role in the creation of international space law. Several major treaties and international agreements address the international community's use of outer space, and the extensive commercial use of space has also led to the creation of private international space law, which consists of transnational contracts that deal with liability, responsibility and procedures for potential disagreements and conflicts between companies and agents. This chapter, however, focuses on public international space law and the existing agreements which outline states' rights and obligations in outer space and the upper levels of the Earth's atmosphere.

For the most part, these treaties were developed in the 1960s and 1970s against the backdrop of the Cold War. Since the end of the Cold War, they have not changed very much. Because existing space-related multilateral agreements do not incorporate the technological and geopolitical changes that have taken place in recent years, there is a real need to kick-start ongoing discussions and debates and to revise international space law. That said, the establishment of a new regulatory regime for outer space will, for a variety of reasons, be challenging in the short to medium term.

The development of international space law

International space law has often been compared to the law of the sea or the law of airspace, since these examples also deal with a strategic sphere in which national sovereignty and borders are difficult to specify and where there is a common interest in keeping such spheres usable and free for peaceful passage. More recently, international environmental law has also been connected to space law, due to the need for a multilateral approach to handling violations and threats.

Adopting a view of space law that correlates to the law of the sea encourages the view of outer space as 'an open access area' which can be used by all and appropriated by none.[72] It also implies that each state can enforce legal rules on its own nationals wherever they are in space. If airspace law were the model for space law, however, it would imply that outer space would be separated into national segments and only allowed to be used with the authorization of the state possessing that segment. It would also mean that each state would be able to enforce the legal rules on every person that operates in its national segment.[73]

The choice of which legal regime to base space law on is a crucial decision, since it has significant implications for space exploration, orbiting objects and satellite reconnaissance. This discussion was initiated long before the first launch of a satellite and most Western international lawyers supported an approach similar to that taken with the law of the seas.[74] This opinion was argued based on physics and technological capabilities. For example, due to the Earth's rotation in orbit around the

Sun, geographic relations between states and particular areas of outer space constantly change. It is also inherently difficult to find appropriate means of securing national borders and excluding others from sovereign areas in space.[75]

The launch of Sputnik by the Soviet Union in 1957 was also the starting point for tangible discussions on space law. The launch showed the international community that there was an urgent need to develop international rules and regulations for behaviour in outer space and to put previous theoretical discussions into practice. Outer space was a new field that was separate from national territory, and when it suddenly became possible to operate in such a sphere it was essential that human activities should be appropriately managed.

Before Sputnik, there was no regulation of space. The launch of Sputnik laid the foundations for the most basic principle for activities in space: that space is free and belongs to mankind without territorial limitations. It has been suggested that the United States was pleased about the Soviet launch because it resolved the dilemma of the overflight of satellites. The Soviet Union could not argue that its own satellite would violate other nation's national territory and, through the launch of Sputnik, the concept of vertical sovereignty was dismissed and the principle of the free passage of satellites was coined. It is also important to note that there were no anti-satellite weapons or similar capabilities at that time, making it difficult to prevent satellites from passing over territory. However, there have been instances where vertical sovereignty has been pushed, most notably in the *Bogotá Declaration*,[76] issued in December 1976 by a number of equatorial states, in which they claimed rights of sovereignty over the cosmic space above their territories. All objects placed in the lucrative geostationary orbit pass over equator, making this an important issue for space-faring nations and corporations. The Declaration regarded geostationary orbit as a natural resource that should be considered to fall to the national territory. The arguments made in this Declaration are still discussed, most notably during each session of COPUOS. Most non-equatorial countries state during these annual meetings that accepted international law, such as the OST, emphasizes the freedom of exploration of space for the benefit of all and that such wording runs counter to the claims of the *Bogotá Declaration*. The declaration has no legal standing and, after opposition from most developed countries, many equatorial countries moved away from the position stated within it.[77] However, the questions raised in the *Bogotá Declaration* relate to unresolved international space law issues, such as the definition and delimitation of outer space.

Shortly after the launch of Sputnik, the General Assembly took the first concrete steps to regulate space. COPUOS was established in 1959, and in 1963 the General Assembly adopted *the Declaration of Legal Principles Governing the Activities of States in the Exploration and Use of Outer Space*.[78] This Declaration is considered to be the basis for the OST, which, together with the four multilateral treaties listed above, forms the treaty-based part of international space law. While space appears to be a very different geographical entity and one in which appropriate regulations are absent, it is important to remember that international law is still applicable. The principle of being subject to international law, including the Charter of the United Nations, was first mentioned in the Declaration of Legal Principles of Outer Space and later codified in the OST. This means, for example, that, with exception of cases in which Article 2 of the United Nations Charter can be invoked, the use of force or the threat of the use of force either in space or towards assets placed in space is a violation of international law and therefore prohibited.

The legal framework of outer space

The main international legal tools dealing with outer space are implemented in a framework of five multilateral treaties. Central to this framework is the OST. The principles of the five treaties listed in the section above 'provide for non-appropriation of outer space by any one country, arms control, the freedom of exploration, liability for damage caused by space objects, the safety and rescue of spacecraft and astronauts, the prevention of harmful interference with space activities and the environment, the notification and registration of space activities, scientific investigation and the exploitation of natural resources in outer space and the settlement of disputes'.[79] The five treaties were adopted by the United Nations General Assembly and today are considered international customary law.

(1) The Partial Test Ban Treaty

Between 1953 and 1958, the Soviet Union, the United Kingdom and the United States conducted a total of 231 atmospheric tests of nuclear weapons.[80] As the arms race and tensions over nuclear weapons between the United States and the Soviet Union reached a fever pitch in the late 1950s and early 1960s, it became increasingly obvious that certain regulations were necessary to prevent widespread nuclear fallout and inadvertent radiation poisoning from the superpowers' ongoing nuclear weapons tests. The nuclear tests raised concerns over radiation levels in the atmosphere and the subsequent risk to human health, and public outcry over the tests grew increasingly loud. In 1958, the Soviet Union unilaterally decided to suspend its atmospheric tests of nuclear weapons. Pressure grew for the United States to take similar actions, and the culmination of this was the 1963 signing of the Partial Test Ban Treaty (PTBT), which prohibits nuclear tests and other nuclear explosions in the atmosphere, underwater or in outer space. By placing limitations on nuclear weapons testing in outer space, the international community took its first concrete step towards active regulation of states' activities in this domain and indirectly affirmed the idea that space should be used for primarily peaceful purposes. The PTBT also shows a common approach to the seas, the atmosphere and outer space and was a clear signal of an international consensus that such areas are commonly owned and that environmental catastrophes in such areas must be avoided.

(2) The Outer Space Treaty

The OST was negotiated in the 1960s, at the height of the space race. It was opened for signature on 27 January 1967 and entered into force nine months later.[81] As of July 2011, 100 states had ratified the Treaty and a further 26 had signed it.[82] The OST is the main treaty regulating activities in outer space. It lays out the core principle guiding all actions in space, that activities in space should enhance the well-being of all countries and all humankind. The OST explicitly prohibits military activities on the Moon and other celestial bodies and provides general guidelines for the peaceful use of outer space. This principle has been confirmed in subsequent United Nations resolutions as well as other discussions on outer space, and is now considered international customary law.

The OST was based on the Antarctic Treaty,[83] which regulates international relations with respect to Antarctica. They have many characteristics in common in that they seek to prevent colonial competition and damage through exploitation, and prohibit the establishment of military bases and fortifications.[84] After the signing of the PTBT, the United States sought to capture the momentum in favour of arms control agreements and pushed for a treaty on the placement of weapons of mass destruction in outer space. In 1966, the United States and the Soviet Union submitted draft treaties, and in September 1966 they reached agreement in negotiations in Geneva.[85]

Article 2 of the Treaty sets out the requirement for peaceful uses of outer space, although it contains no definition of the concept. There is an extensive legal debate among international jurists over whether '*peaceful uses*' should mean 'non-aggressive' or 'non-military'.[86] Outer space is evidently militarized, due to the reliance on space for conducting military operations. Thus state practice shows that the international community seems to have accepted 'non-aggressive' as an appropriate interpretation of peaceful purposes, but the ambiguity remains. Article 4 of the OST bans weapons of mass destruction in space, but it does not prohibit missile-borne weapons of mass destruction from transiting space. Nor does it ban other types of weapons from being placed in orbit or being used to attack targets in space or on Earth. For example, nuclear warheads on Ballistic Missile Defense (BMD) interceptors launched from the ground are not prohibited, and 'nor is the sub-orbital transit of outer space by nuclear warheads on ballistic missiles'.[87] The Treaty does not prohibit the use of space for military purposes. Because the OST does not make any stipulations on the use of conventional weapons in outer space, there is still room for states to use ground, sea or air launched or space-based anti-missile weapons.[88] Thus, while the Treaty strongly advocates peaceful uses of outer space, it does not definitively address the issue of weaponizing space or using space for military purposes. Not surprisingly, this shortcoming is among the most heated areas of debate in contemporary space security negotiations.[89]

Another key shortcoming of the Treaty is the fact that it does not define weapons of mass destruction, leaving at least some room for flexible interpretation. The treaty is also silent on a definition of outer space. The treaty only mentions that the Moon and other celestial bodies are included in outer space, but nothing about where outer space borders airspace. Nor does it deal with the question of assets in space that have been brought back to Earth: are they under national ownership when they enter airspace or will they always be for all of mankind? This might become a pressing issue if the technology for asteroid mining develops further, and resources from space can be used here on Earth. Another limit of the OST is that it does not address the question of implementation procedures and verification mechanisms.[90] It is important to keep in mind the geopolitical complications and the competition in space that existed during the negotiation of the OST. It could be argued that the drafters avoided specific issues and clarifying definitions in order to reach a compromise between all the nations concerned.[91] It has been argued that the aims and principles of the treaty are its most important contribution to space law rather than specific definitions. It has even been called 'a Constitution for Outer Space'.[92]

The OST does not include any provisions for review, and no amendments have been made to address any of the above gaps. In June 2006, the International Weapons of Mass Destruction Commission presented its report to the United Nations Secretary-General. The report makes proposals on how to reduce the dangers of weapons of mass destruction and included a recommendation concerning the flaws of the OST: '[a] Review Conference of the Outer Space Treaty to mark its 40th year in force should be held in 2007. It should address the need to strengthen the treaty and extend its scope. A Special Coordinator should be appointed to facilitate ratifications and liaise with non-parties about the reinforcement of the treaty-based space security regime'.[93] The geopolitical situation has changed since the Treaty was concluded, there have been enormous advances in technology and the number of actors with access to space has increased. The progress of science and our increasing dependence on space make it necessary to improve the legal framework of space security. However, a review of the treaty could be complicated, not least because provisions for one are not included in the original wording. Any amendment would require states parties to open up the treaty for renegotiation and would risk the current status of the legislation within it. Despite its evident flaws, loopholes and gaps, the OST remains the main legal regime for space activities.

(3) The Astronaut Rescue Agreement

The Rescue Agreement was adopted by the General Assembly in 1968.[94] It was negotiated by the Legal Committee of COPUOS. The Agreement was the first to be concluded after the OST, and further elaborated on Article 5 of that Treaty, which deals with the safety and rescue of astronauts. Article 5 of the OST states that 'States Parties to the Treaty shall regard astronauts as envoys of mankind in outer space and shall render to them all possible assistance in the event of accident, distress, or emergency landing on the territory of another State Party or on the high seas.'[95] With ten articles, the Rescue Agreement sets a legal framework for dealing with astronauts and accidents in space. The political will to conclude this treaty was enhanced by two tragic events in the 1960s. The first happened in January 1967, when a fire broke out on the US spacecraft Apollo I in which three astronauts died.[96] In the second event, some three months later, a Soviet cosmonaut was killed when the Soyuz I spacecraft crashed on re-entry into the Earth's atmosphere.[97] While the Rescue Agreement would not have been able to save these lives, the incidents highlighted the need for cooperation on rescue attempts.

The Rescue Agreement makes a clear distinction between the personnel in a spacecraft and the spacecraft itself, and differs from many other multilateral treaties in that it requires states parties to inform the United Nations and its Secretary-General about steps taken and their progress. It is claimed that this originated from the military applications of spacecraft during the Cold War. Both sides wanted the United Nations to be notified in cases where military space technologies or personnel crashed on foreign territory.

The Agreement reflects the technology level and the geopolitical situation of the 1960s and, while international organizations responsible for space launches can become a party to the Agreement, it does not properly address the role of commercial companies in space. New conditions and new problems have emerged which were not adequately dealt with in the 1960s. There is an argument for creating a new regime for the safety of space travellers today that can deal with personnel stationed in space permanently and potentially also commercial space travellers such as space tourists.[98] Space-faring capacities have developed and new conditions have been put in place since the Agreement was signed. The Rescue Agreement has been called dated and incapable of dealing with today's realities. There are several ways that the Agreement could be strengthened, such as adding a provision on dispute settlement, broadening it and further defining the term 'launching authority', since today's space launching entities can be multinational corporations, collaborations between nations or multinational organizations.[99]

(4) The Liability Convention

The Convention on International Liability for Damage Caused by Space Objects, often called the *'Liability Convention'*,[100] establishes liability for space accidents and collisions. The main issue, that states must bear responsibility for space-related damage, is one of the most complex issues in space law. The Convention was negotiated by the Legal Committee of COPUOS,[101] and took almost ten years to conclude.[102] The Convention complements the OST and is broad in scope with the aim of capturing all possible types of damage that are potentially harmful in space. Under the Convention, the launching state is absolutely liable for surface damage to aircraft in flight by a space object. It requires that if a space object owned by one state causes damage to objects belonging to another state – whether those objects are on the surface of the Earth or in the air – the state whose space object caused the damage is liable and must compensate the other state accordingly.[103] Currently,

90 states have ratified the Convention and an additional 23 are signatories.[104] The Liability Convention was pursued because of the expanding number of activities in space and the potential economic and political consequences of disputes between space actors. Since the number of actors is increasing, both states and commercial entities, the Liability Convention faces some serious challenges.

Article 2 of the Convention is one of its most central. It states that '[a] launching State shall be absolutely liable to pay compensation for damage caused by its space object on the surface of the Earth or to aircraft in flight.'[105] Article 4 clarifies that, according to the requirements of the Convention, only the launching state will be responsible for the damage. Considering the number of private actors in space that are not included in the Convention, this leads to some complications as to who is responsible for such actors. According to Article 1 (c), '[t]he term "launching State" means: (i) [a] State which launches or procures the launching of a space object; (ii) [a] State from whose territory or facility a space object is launched.'[106] This means that it is not the launching company that will be held liable, but the state in which the launch took place. At a time when countries, space agencies and commercial entities are expanding their cooperation, this might become a source of conflict in the future.

There are a number of additional complications with the Convention, such as insufficient definition of the term 'space object' in Article 2, as well as other terms such as 'fault' and 'gross negligence'. This lack of or vagueness in definitions could create serious loopholes in the future.[107] The Convention also lacks any clarification about damages. Does it, for example, specify indirect damages or only direct? It also contains a complicated exception to the principles of compensation in Article 7, which states that

> [t]he provisions of this Convention shall not apply to damage caused by a space object of a launching State to: (a) nationals of that launching State; (b) foreign nationals during such time as they are participating in the operation of that space object from the time of its launching or at any stage thereafter until its descent, or during such time as they are in the immediate vicinity of a planned launching or recovery area as the result of an invitation by that launching State.[108]

It is these two categories of people that might be considered to run highest risk of an accident.[109] Because of the nationalistic approach to space programmes during the time of negotiations, the Convention does not adequately deal with cases of multilateral space ventures, such as today's multinational corporations, organizations or collaborations between national authorities. Like the Rescue Agreement, the concept of launching state is in need of modification and adjustment in order to reflect today's reality. New challenges are emerging because the number of objects in space and of debris-creating events is increasing. For example, would the Chinese Government be liable for damage caused by the remaining debris from its ASAT test in 2007? The Liability Convention has serious flaws in the post–Cold War reality. There are more and more commercial entities in space and the Convention should be adjusted accordingly. This leads to the same problem as with the OST. The Convention has no provision for amendment and would therefore have to be reopened for negotiation. Considering that the original Convention took ten years to negotiate, such a reopening could be difficult to achieve.

(5) The Registration Convention

The Convention on the Registration of Objects Launched into Outer Space[110] requires member states to inform the United Nations of their space launches, the general function of the launched objects and the objects' orbital parameters. This is the third treaty based on the provisions of the OST, and the Convention has been ratified by 55 states and has four other signatories. International groups such as the European Space Agency and the European Organization for the Exploitation of Meteorological Satellites have also publicly stated that they will accept the rights and obligations outlined in the Convention.[111] It is Article 13 of the OST that has been elaborated into a complete Treaty. It states that 'the provisions of this Treaty shall apply to the activities of States Parties to the Treaty in the exploration and use of outer space, including the Moon and other celestial bodies, whether such activities are carried on by a single State Party to the Treaty or jointly with other States, including cases where they are carried on within the framework of international intergovernmental organizations'.[112] Article 2 of the Convention sets out the general obligations for states parties concerning the procedure of the registration of objects and Article 3 calls on the Secretary-General to maintain the register of all registered objects. It also clarifies that '[t]here shall be full and open access to the information in this Register'.[113]

The Registration Convention provides that the launching state should furnish the United Nations, as soon as practicable, with the following information concerning each space object:

- name of launching state;
- an appropriate designator of the space object or its registration number;
- date and territory or location of launch;
- basic orbital parameters, including nodal period, inclination, apogee and perigee;
- general function of the space object.[114]

Registration of space objects is seen as an important part of preserving space for peaceful purposes since it minimizes the likelihood, and suspicion, of the placement of weapons of mass destruction or other hostile objects in space. However, the issue of the 'general function' of a space object is controversial, since most countries are unwilling to provide any information about the military objectives of their space assets. While the registry of space objects, administrated by OOSA, is a confidence-building measure and a transparency enhancer, many states do not trust each other enough. OOSA provides an open database with information available on all objects registered since the launch of the registry.[115] For national security reasons, further information about space objects is difficult to obtain. The Registration Convention includes a provision for revision. Article 10 states that '[t]en years after the entry into force of this Convention, the question of the review of the Convention shall be included in the provisional agenda of the United Nations General Assembly.'[116] Despite some arguments that Article 4 should be changed to include an obligation for registering states to provide information about the presence of armed systems on board, the General Assembly decided that there was no need to amend or review the Treaty.[117]

(6) The Moon Agreement

The Moon has always been fascinating to humans. The US Moon landing in 1969 and the discovery of natural resources there increased incentives for a more specific regulation of activities on the Moon. The 1979 Moon Agreement reaffirmed many of the principles of the OST, emphasizing that the Moon and other celestial bodies can only be used for peaceful purposes.[118] It also added a prohibition of '[a]ny threat or use of force or any other hostile act or threat of hostile act on the Moon.'[119] The Agreement made some new contributions to international space law, such as on the exploitation and preservation of the Moon, which should be carried out through mutual assistance, and on different forms of international cooperation.[120] The Agreement expands the concept of outer space being 'the province of mankind', as is stated in the OST, and declares that the Moon and all its resources are 'the common heritage of mankind' and that its resources should be collectively exploited and shared among the entire international community.[121] It also establishes a legal regime to manage the exploration of the Moon's resources. Despite lengthy negotiations and the fact that the text was originally adopted by consensus in the General Assembly, only 13 countries have ratified the 1979 Moon Treaty, and none of the major space powers has done so.[122] It entered into force on the basis of its ratifications, but it cannot be regarded as a success since neither the United States nor Russia supports it. Disagreements mainly concern the question of natural resources on the Moon and the establishment of the international regime to deal with such exploitation in Article 11.[123] Article 11, paragraph 5, states that 'States Parties to this Agreement hereby undertake to establish an international regime, including appropriate procedures, to govern the exploitation of the natural resources of the Moon as such exploitation is about to become feasible.'[124] Paragraph 7 in the same article continues:

> The main purposes of the international regime to be established shall include: (a) the orderly and safe development of the natural resources of the Moon; (b) the rational management of those resources; (c) the expansion of opportunities in the use of those resources; (d) an equitable sharing by all States Parties in the benefits derived from those resources, whereby the interests and needs of the developing countries, as well as the efforts of those countries which have contributed either directly or indirectly to the exploration of the Moon, shall be given special consideration.[125]

While these provisions reflect similar efforts in the Convention of the Law of the Sea, the international regime that would govern exploitation of the natural resources of the Moon has not yet become a reality. Since this relative failure of the 1979 Treaty, the United Nations system has had trouble establishing itself as a genuine driving force for further development.

(7) Other agreements, resolutions and transparency and confidence-building measures

Treaties play an important role in developing space law. However, since the conclusion of the Moon Agreement, no further treaties have been negotiated in COPUOS. The development of customary law through state practice and soft law commitments is increasing in importance. While such measures are not legally binding, they do provide some steps forward to further institutionalize cooperation and regulation. It is also important to note that once such soft law agreements mature into customary international law, they gain legally binding force. Since the launch of Sputnik, the General Assembly resolutions and declarations have been the most important ways of creating international customary law.

Since the 1950s, the annual session of the General Assembly has included *international cooperation in the peaceful uses of outer space* on its agenda. Resolutions are adopted and the work of COPUOS is reviewed under this agenda item. The General Assembly adopts an annual resolution on *the Peaceful Uses of Outer Space,* which includes a reaffirmation of the principle that space exploration is for the common interests of mankind, the importance of international cooperation in developing the rule of law in space as well as calls for member states to adhere to the *Space Debris Mitigation Guidelines.* For many years Russia has presented an annual resolution to the General Assembly's First Committee on the *Prevention of an Arms Race in Outer Space.*[126] This resolution consists of calls on the international community to commence negotiations in the CD on a treaty that would ban weapons in space, such as the proposed *Prevention of the Placement of Weapons in Outer Space Treaty.* This resolution has been adopted each year by a majority vote, with only the United States voting against and Israel the only abstaining state. However, in 2009, the United States, which had voted against this resolution since 2005, switched to an abstention. Another annual resolution proposed by Russia as of 2005 is on *Transparency and confidence-building measures in outer space activities.*[127] This initiative led to the setting up of a Group of Governmental Experts that will begin its work in 2012 under the auspices of the United Nations. Its conclusions about the EU Code of Conduct and the PPWT will provide important insights into governance issues and diplomatic initiatives over the next couple of years.

A collection of principles has also been adopted by the United Nations. These principles are considered to be a special category of resolutions, starting with the first legal regulation of outer space mentioned above, the *Declaration of Legal Principles Governing the Activities of States in the Exploration and Use of Outer Space*. This Declaration was the starting point for the OST and the 'first comprehensive international intergovernmental document reflecting the position of the world community as a whole concerning general principles by which States should be guided in the exploration and use of outer space'.[128] While not legally binding at the time of its adoption, this document contained most of the principles which would later be incorporated into the OST. Since the Treaty is not universal, the Declaration continues to fulfil a function for non-parties to that treaty. It is argued that this Declaration contains customary international space law, due to its unanimous adoption and the intention of the General Assembly to make it a document of strong significance with a clear expectation that members of the international community would abide by it.[129] This means that violations of the Declaration made by non-parties to the OST would nevertheless be regarded as a breach of universally recognized international law.

The second set of principles is the *Principles of Direct Broadcasting by Satellites*. These negotiations took over ten years and contain ten sections dealing with broadcasting principles, such as applicability to international law, the peaceful settlement of disputes and state responsibilities.[130] COPUOS member states could not reach consensus when negotiating these principles, leading to their adoption through a majority vote in the General Assembly. Since a few countries, some of which are very active in the exploration and use of outer space, did not support these principles, they cannot be regarded as customary international space law.

The *Principles on Remote Sensing*,[131] adopted in 1986 through the unanimous approval of the COPUOS member states, serve as general guidelines for conducting remote sensing from outer space. They are not binding, but are adhered to by states. However, no dispute or conflict has arisen concerning this area and no case law clarifies the legal application of the guidelines. In 1992, the *Principles on Nuclear Power Sources*[132] were adopted, also by a unanimous decision by COPUOS. The issue of nuclear power sources in space has been a controversial one since a Soviet nuclear powered satellite, Cosmos 954, crashed in a remote area of Canada, contaminating it with radioactive debris.[133] While it is fortunate that the satellite entered the atmosphere in an unpopulated area, it highlighted the complex subject of nuclear powered satellites. The principles contain detailed guidelines for safe use, including such tasks as notification of re-entry, assistance to states, liability and compensation, as well as settlement of disputes.[134] Although these principles are not legally binding, they carry moral and political weight, mainly due to the sensitive nature of nuclear energy. The final set of principles is the *Declaration on Outer Space Benefits*.[135] Adopted unanimously in 1996, the declaration aims to ensure access by all countries to the benefits of space technologies. Some developing countries argued that there was a lack of legal efforts to ensure that space was used for the benefits of all countries, particularly considering the great gap in technological capability between the world's states. However, the declaration has been of limited use and does not consist of any specific measures to combat such problems.

One of the reasons for choosing to pursue such declarations of principle instead of treaties is that the area of international space law is relatively new. It is possible that some states regard the domain of space as too new to consider establishing international law to regulate it and instead prefer to develop principles and customs at a slower pace through state practice. It is also true that the science and technology that enable space exploration are areas where rapid progress is very common. There might be some reluctance to adopt treaties when the possibilities of space exploration are constantly changing and developing. Given the outdated provisions in the five space treaties, we can see that this is indeed a problem, since treaty obligations are difficult to change without reopening negotiations.

Outside of United Nations, there are several arms control treaties and agreements, as well as soft law initiatives that are also relevant to states' activities in outer space. These agreements include the *Strategic Arms Limitations Talks* (SALT I and II),[136] the *Strategic Arms Reductions Treaty* (START)[137] and the *Intermediate Range Nuclear Weapons Treaty*.[138]

While not part of any international space law regime, the *Missile Technology Control Regime* (MTCR) regulates missile technology and has a space dimension due to the similarities between missile technologies and launch rockets. The MTCR is an informal and voluntary association of states with the purpose of preventing the proliferation of delivery systems of weapons of mass destruction by coordinating national export licensing.[139] The MTCR has a current membership of 34 countries.[140] It was initiated in the late 1970s by the United States, which was concerned about events such as South Korea's ballistic missile test in 1978, India's test of its SLV-3 in 1980 and Iraq's attempt to purchase rocket stages from Italy in 1979.[141]

The MTCR aims to control complete rocket systems, such as ballistic missiles, space launch vehicles and sounding rockets, as well as unmanned air vehicles (UAVs), such as cruise missiles, drones and remotely piloted vehicles. The regime is based on a set of common export policy guidelines, called the MTCR Guidelines.[142] The member states control the export of an agreed list of items found in the MTCR Equipment, Software and Technology Annex.[143]

The MTCR has proved to be an efficient instrument for combating the proliferation of missile technology, and it has brought technology supplying countries closer together through standardization and cooperation on export policies. It has also raised awareness of accountability over the export of critical technologies.[144] Like all initiatives, however, there are some problems. It is not a treaty-based agreement, so all the obligations and rights under the MTCR are voluntary and there are neither enforcement procedures nor non-compliance mechanisms. Violations of the guidelines have been countered in different ways in the past. The United States, as the most ardent MTCR participant, backs the implementation of the MTCR by a sanctions law, whereas other states have taken a more low-profile approach. For example, in January 1992, Russia concluded an agreement with the Indian Space Research Organization (ISRO), through which Russia would provide India with rocket engines and related technology. While Russia claimed that India intended to use this technology for peaceful space applications, the United States argued that it countered the MTCR and imposed sanctions against the Russian space and missile industry and the Indian Space Research Organization.[145] In 1993, China was accused of violating the MTCR guidelines. Although not a member, China made a unilateral declaration in 1992 that it would abide by the MTCR.[146] However, China refused to embrace some changes in the guidelines, and when China sold components for missiles to Pakistan in 1993 the United States responded with sanctions on technology exports.[147]

It is argued that the guidelines are weak and unclear, which has led to inconsistent implementation by member states.[148] In addition, since the membership is limited, one of the main challenges to the MTCR is that non-members, such as North Korea, for example, can continue to share and transfer missile-related technology. The Iranian space launcher, which was tested in April 2009, was built with North Korean technology. The MTCR is an evolving regime and its member states hold annual meetings in which expansion of the membership and discussions on making the regime more effective take place.

The *Hague Code of Conduct* (HCoC) was adopted at a conference in The Hague in November 2002. It aims to prevent the proliferation of ballistic missiles and has more members than the MTCR.[149] The HCoC was initiated to supplement the MTCR, and membership is open to all states. There is no formal international treaty which prohibits the development, acquisition or export of ballistic missiles so it has been argued that the HCoC addresses a gap in the international non-proliferation regime.[150] While the MTCR restricts the export of missiles, the HCoC calls on subscribing states to show restraint in their national missile programmes.

The HCoC consists of a set of general principles and commitments and a list of confidence-building measures. The HCoC has established an international norm against the proliferation of ballistic missiles. It aims '[t]o curb and prevent the proliferation of Ballistic Missiles capable of delivering weapons of mass destruction, both at a global and regional level, through multilateral, bilateral and national endeavours'.[151] It is by far the most advanced initiative taken by the international community to regulate ballistic missiles. Like the MTCR, the HCoC is not a legally binding treaty, and there is no inspection regime to assure compliance and no procedure for dealing with violations.

However, the HCoC does include some political commitments. All member states have agreed to abide by the measures set out in its Article 4. These transparency and confidence-building measures consist of a declaration of ballistic missile policies, information on the number and generic class of ballistic missiles launched, an outline of Space Launch Vehicle policies and test sites and invitations to international observers to launch sites as well as exchanges of pre-launch notifications on ballistic missile and Space Launch Vehicle launches and test flights.[152]

While membership of HCoC is by no means universal (North Korea, Iran, India and Pakistan have not signed it), it provides a basis for further deliberations. Apart from the lack of legally binding requirements, the HCoC has other shortcomings. Some states have not reported their missile or space launch programmes, for example. It has also been argued that the HCoC is too limited and should be expanded to cover other delivery systems, such as cruise missiles and UAVs. These missiles are more suitable for delivering biological and chemical weapons. The HCoC also lacks a secretariat, although Austria was appointed the administrative central contact coordinating information exchange under the HCoC. It is argued that the HCoC is weaker than the MCTR because it does not mention cruise missiles and has weaker control of space launch vehicles. This could risk encouraging countries to 'choose' the weakest regime.[153]

In parallel, a number of states proposed codes of conduct for outer space activities, as an alternative approach to TCBMs. The European Union adopted a Code of Conduct for Outer Space activities in 2008 and presented it at the CD in 2009. Canada has also proposed an alternative TCBMs initiative, standing between the EU Code and the PPWT. Taking the form of a 'declaration of soft legal principles', it was presented to the CD and codified as a CD document in June 2009.[154]

The need for a new approach to the international regulation of outer space

The increasing number of space-faring nations, major advances in space technologies and the spectacular rise of the commercial space industry must be taken into account when discussing the future of space. None of the five main treaties defines what a space object is or what a spacecraft is, and they were all drafted with states in mind rather than the commercial space sector we see today. Space requires international laws, regulations and operational guidelines that can successfully govern 'its use to enhance commercial and scientific uses, avoid and settle conflicts, and limit debris production'.[155] There are currently few international mechanisms for managing *traffic* in space, coordinating the placement of satellites in Earth's orbit and dealing with the proliferation of space debris from spent boosters and defunct satellites.[156] However, a number of recent incidents in space have shown that successfully addressing such challenges requires a new approach to space security and international space regulation.

One of the key objectives of certain states is an international treaty banning space weapons. The problem with the OST and the other legal documents regulating various issues related to outer space is that none clarifies or defines what the 'peaceful use of outer space' really means. For example, the United States and the Soviet Union focused on the potential for military, communications and intelligence uses in space, and came to understand 'non-aggressive' as leaving room for the potential militarization of space. Most analysts agree that the costs related to the weaponization of space would outweigh any potential benefits[157] and the General Assembly continues to call on the CD 'to establish an ad hoc committee to draft a multilateral treaty to prevent an arms race in outer space'.[158]

Recently China and Russia have been leading the call for a treaty that would ban the use of all weapons in outer space. These two space powers fear the prospect of a space arms race, and they would also like to prevent the United States from deploying an anti-ballistic missile defence system. However, any further progress towards a consensus on a set of legal measures for sustained, secure and safe use of outer space requires a more global and less state-centric approach to space security. While the value and, consequently, the probability of states placing weapons in space is not clear, other, less expensive and technically easier but nevertheless harmful, activities such as jamming and cyber attacks need to be urgently addressed. Fostered by globalization and the shift in the security environment, these threats have recently grown in importance. Mostly perpetrated by independent groups rather than states, these attacks aim to deny access to space-based systems, are cheap and require no particular infrastructure. Yet, they can be serious enough to damage the systems modern societies rely on and their perpetrators remain difficult to identify. Today, there are no rules to mitigate this type of risk and no criminal law to identify and sanction the perpetrators of such actions. There is also an urgent need for action on space debris. Recent ASAT tests by the United States and China showed that this is a serious problem that could easily get out of control, leading to a reduced capacity to act in space. COPUOS has adopted a set of guidelines on debris mitigation, but it is not certain that the guidelines can provide enough legal commitments from states in order to prevent increased levels of debris.

There have been discussions about a ban on the testing of ASAT weapons,[159] since additional testing of such weapons would be devastating for the space environment. While such a ban would not prohibit the use or existence of ASAT weapons, it would be a starting point and might put limits on states with such capabilities. It would also be a useful first step towards greater treaty-based regulation of space and be easier to achieve than a general treaty on space weapons. However, there are complications with such proposals. A situation similar to that of the Nuclear Non-proliferation Treaty could easily develop, where two sides are created: those that have weapons legitimately and those that do not. Many countries would protest if the United States, China and Russia, for example, were legally allowed to possess an additional powerful weapon that others were not allowed. Such a treaty could also trigger the proliferation of ASAT testing before agreement to any negotiations, since some countries would not want to 'miss the bus' and risk being left out of another weapons treaty. Therefore, it is important that the international community includes ground-based ASAT testing in the future space regulation framework, as more ASAT tests and, by extension, more space debris will result from the inevitable competition between states.

Another issue that should get more attention is the establishment of international standards and verification measures. None of the five treaties contains a set of standards or procedures against which the implementation of these treaties can be verified. There is also a need to develop space law in the area of the commercial use of space. The commercial sector is rapidly expanding and the legal implications for issues such as property, liability, definitions of launching state, insurance and other critical issues can create serious problems in cases of conflict or disagreement. Recent developments in the area of commercial space tourism and private manned space flights suggest that future space law needs to take the increased civil presence in space into consideration.

There are also many divisions and problems relating to the institutions dealing with space. There is a jurisdictional problem between the different agencies which have been mandated to deal with space security. While the CD is strictly focused on arms control in space, COPUOS has a distinct mandate which deals with the peaceful uses of space. The ITU addresses technical issues concerning the radio frequency spectrum and orbital slots, while the IADC deals solely with space debris mitigation. OOSA mainly works as a coordinator and secretariat for COPUOS and has no autonomous decision-making powers. These divisions of labour can make the complex area of space security even more complicated. For example, the issue of space is marginal to the CD. It only looks at a small part of the puzzle. It is an arms control body, which discusses weapons in space but not the economics, scientific development, military uses, civilian uses and commercial uses. The CD does not discuss debris or radio frequency technologies. Nor does it deal with campaigns for the convergence of space technology. This division of labour might seem logical at first, but the inherently dual-use characteristics of space technology and the difficulties of separating areas of use create problems with defining areas of work within these organizations.

Despite its specific technical area, the decisions and regulations of the ITU are strongly debated for political reasons. COPUOS and the CD have launched a limited collaboration. The connection between experts at the IACD and the diplomats in COPUOS is also a very important tool for successful negotiations. The national space agencies in the IACD have had a clear interest from the early days in avoiding debris and, when the time was ripe, that interest also grew at the governmental level. The work on the IACD guidelines has provided a basis for negotiations in COPUOS, despite the different mandates and different rules. This clearly shows that discussions in different bodies and in different settings can spill over and affect other organizations. It also shows that experts play an essential role in resolving technical problems as well as identifying common interests outside the political arena.

The development of a new legal space instrument requires a *holistic approach* to space treaty making. The OST created a foundation for the global regulation of outer space, but it covers only celestial bodies and not space in its totality. Solid verification provisions are crucial. The elements of a treaty to prohibit the weaponization of space would need to include a ban on the deployment of any weapons in space, and a ban on the testing and use of Earth-or space-based anti-satellite weapons. It must also include cyber criminality and cover privatization and space traffic management. The existing regulations dealing with space are separated into two areas: those that deal with the peaceful uses of space and those that consider military and weapons-related issues. There is a clear lack of an overarching framework that is adequate for dealing with the challenges we face today. There is a need for a larger space regime, which would address the full range of space-related issues. Such a holistic and globally accepted space treaty would enhance space security and stability. However, although this legal framework would be a huge achievement, the development of such a treaty is very unlikely in the near-future.[160] As argued by Theresa Hitchens, '[f]rom the struggle in all three fora, [COPUOS, ITU and CD] it is clear there is a widespread reluctance among states to enact new legal restraints on space activities in any domain. On the other hand, it is apparent that there is a gathering impetus for "soft law" action to mitigate the twin probems of space safety and security.'[161]

In fact, the adoption of the guidelines for debris mitigation in COPUOS has shown that through technical expertise and soft law approaches, a shared vision of enhancing general security for all can become a reality. While soft law and TCBMs such as codes of conduct might not be a watertight regime for space security, it is important to remember that international obligations arise not only from treaties but, according to Article 38 of the Statute of the International Court of Justice, also from general practice law such as international custom. Furthermore, soft law tools can 'contribute to promoting openness and to building confidence in outer space activities, so as to ensure transparency in the use of outer space, to avoid collision or interference, to avoid conflicts and to prevent the deployment of weapons in space'.[162] Nevertheless, soft law instruments are not substitutes for arms control and disarmament measures. They can serve as a first step towards the development of a legally binding treaty and can be incorporated later in the future regulation. The international practice of the past 30 years has made it clear that the preference of major space powers is for a *rules of the road* approach, which could serve as a starting point for future space governance discussions.[163]

Meta-geopolitics and its implications for outer space in the twenty-first century

Global policymakers face a fast-moving, multipolar, globalized and interconnected world. In this context, access to and the use of space have gained strategic and political importance, and space has become one of the core elements of international power. Although developed from existing geopolitical theories, such as classical geopolitics and critical geopolitics, the *meta*-geopolitics framework goes beyond these traditional approaches to offer an alternative, all-encompassing framework that helps clarify power relations and, in this context, the correlation between geopolitics in space and their impact on international configurations and global security on Earth.

Meta-geopolitics identifies seven capacities, which are used to gauge a state's ability to project power. This multidimensional approach enables a comparative assessment of space powers. It helps to predict the future geopolitical postures of space powers, while taking account of today's sometimes diffuse and shifting security threats. This analysis offers a nuanced picture of geopolitics in the twenty-first century and provides some clarity on the relationship between our terrestrial globalized world and space. While I recognize that power predispositions on Earth are critical to a state's presence in space and influence its space policy and programme, the space capabilities it acquires, in turn, enhance or weaken its power capacities on Earth and therefore have the potential to alter its overall geopolitical posture. Furthermore, I argue that the framework of *meta*-geopolitics, although initially devised to gauge a state's power, can also be applicable to transnational non-state actors.

After setting out the overall theoretical foundations of the book, I analysed current and future space technologies in order to explain their possible strategic applications. Space power cannot be exerted efficiently unless significant political will is demonstrated and attendant national strategies and policies developed.[164] Possessing an important fleet of satellites is not enough to project power, but this hardware combined with political will, as well as geographic and socio-economic elements, enables states to develop and exert space power internationally. A space programme is a symbol of a great nation. Its spin-offs help the state to improve its capacity and enhance its standing as a global power. Many space programmes are motivated not only by military aims, but also by other aims, such as economic and technological growth, which contribute to the development and independence of the country. The increasing number of states and non-state actors willing to invest enormous amounts in space research shows a clear belief that successful space programmes can enhance power and profit. The analysis of the *meta*-geopolitics of space-power carried out in Chapter 5 shows that no nation controls access to space today, as the United States and Russia did by default during much of the Cold War. While the United States and Russia, together with the European Union (EU) and China, are still the major players in space, new countries and private actors are increasing their presence. The growing role of the private sector has led to the emergence of the global space industry as a new space actor.

However, increasing economic and military reliance on space assets, coupled with the multiple strategic advantages that come with advances in space technologies, have also created new vulnerabilities and raised questions about the sustainability of such power in space. In line with my definition of space power, I argue that the protection of assets in space is essential for sustaining power on Earth. The space environment has profoundly changed since the Cold War, with more satellites in space, more countries operating them and non-traditional actors also investing in space. More dual-use applications are being developed and space is playing a more important role in 'force enhancement' in contemporary conflicts.[165] At the same time, countries have been developing techniques to deny an adversary the use of its space assets. Increased activity in space has led to demonstrations of anti-satellite (ASAT) capabilities, increased collision risks between satellites and a significant increase in hazardous space debris and electromagnetic interference. All this makes it clear that the 'space environment needs more protection, that satellites face growing risks, and that space activities may be a potential source of mistrust and tension'.[166]

Terrestrial geopolitical power struggles have been transplanted to the space realm, and space power will continue to change power dynamics between states on Earth. Hence, space security cannot be separated from security on Earth. Given this close correlation between the two realms, I argued that space security must be placed in a wider context as part of *global security*, in which threats are no longer only of national or military concern, through the utilization of the '*multi-sum security principle*'. While the role of states is to continue to tackle challenges related to their territorial integrity and the physical safety of their citizens, they must look beyond their national borders and military capacities in order to effectively protect and advance their interests.[167]

The *meta*-geopolitics analysis demonstrates that commercial, scientific and technological cooperation exist between the space actors, as there are obvious common interests and significant gains for all parties involved. For example, the successful completion of the ISS will be highly significant for all future international cooperation projects, and it is important that the benefits of the ISS are used to the fullest by all participating states. However, when matters of national security are at stake, this raises obstacles to increased cooperation and an enhanced formalization of agreements. The debris mitigation guidelines discussed in Chapter 4, and further below, are an important step towards enhancing security in space, but these guidelines did little to prevent ASAT tests by China and the United States. Nor did the guidelines prevent the collision between a US and a Russian satellite in January 2009. While it is true that debris mitigation guidelines make subsequent testing by China, Russia or the United States less likely, it is still past time for the international community to increase its efforts and to better prioritize important issues such as space situational awareness, space sustainability and the safety of assets and astronauts in space.

International cooperation in space is one of the most important tools for making progress on security issues, since the future requires a multilateral and global view of space endeavours. It would also make better use of resources, avoiding overlapping projects and creating a more efficient use of funds. Improved international cooperation has clear benefits, such as financial efficiency, industrial development and enhanced access to knowledge and additional skills, as well as the potential to develop specialized and niche capacities. International cooperation is sometimes complex and difficult but it must be encouraged and developed further if we want to continue to conduct challenging and ground-breaking space missions in the future. Cooperation in space between the United States, the EU, Russia, China and other major players in world politics could serve as a powerful model for the entire world. It would provide evidence of the increased benefits available from sharing experience, resources, technologies and other vital components, and provide a basis for strengthened and more transparent relationships among the international community. Yet, although the case for international space cooperation is strong the obstacles and challenges to cooperation are significant as well. Enhancing international space regulation requires states to abandon their nation-centred approach to space security. However, improvements even in such sensitive fields as the military use of space and space security are possible if space is truly treated as a *common good*.

Two paradigms for space governance

As is demonstrated in the chapters above, space is an important component of societal change and global peace and security. All the actors involved in space policies, such as the governments of space-faring nations, international organizations, scientists and astronauts, the private sector, civil society and the academic community, must strive for the creation of a safe and sustainable space environment for future generations. To do so, all space actors must work together to mitigate the risks in space, foster the coordination of space activities to optimize the utilization of space for all and finally 'prevent space activities from increasing tensions between countries that could lead to arms buildups and conflicts'.[168] This section discusses the relevance to space of two paradigms: one, the *'multi-sum security principle'*, on global security; and the other, *'symbiotic realism'*, on international relations. The chapter concludes with some concrete policy recommendations.

The *'multi-sum security principle'*: Relevance to space governance

According to the *'multi-sum security principle'*, discussed in Chapters 2 and 4,

> [i]n a globalized world, security can no longer be thought of as a zero-sum game involving states alone. Global security, instead, has five dimensions that include human, environmental, national, transnational and transcultural security, and, therefore, global security and the security of any state or culture cannot be achieved without good governance at all levels that guarantees security through justice for all individuals, states and cultures.[169]

The first dimension suggested by the *'multi-sum security principle'* is human security, a concept that makes the principle referent object the individual, not the state. The second dimension is environmental security, which includes issues such as climate change, global warming, and access to resources as key parts of global security and stability. The third substrate is national security, which is linked to the state's monopoly over the legitimate use of force in a given territory and is a substrate that emphasizes the military and policing components of security. The fourth component is transnational security, which I define as involving the globalization-mediated security of states against transnational threats such as organized crime, terrorism and human trafficking. Lastly, I refer to the integrity and synergism of diverse cultural and sub-cultural forms as transcultural security. This multifaceted security framework advocates cooperative interaction between states and peaceful coexistence between cultural groups. This principle also emphasizes equity and justice as a central part of global security. As a consequence, all five dimensions of security need to be addressed in order to provide just and sustainable global security. Simply considering one of the five dimensions will not create the appropriate conditions for a sustainable security environment.

Given the significance of space to global security, unilateral actions in space motivated by national security concerns, such as ASAT testing or unwillingness to ban the weaponization of space, cannot improve global security but only threaten it. The space debris resulting from the action of one country puts at risk its own satellites and those of all the other actors at the same altitude, thus affecting the five dimensions of security for all of them. Furthermore, such unilateral actions create suspicion and mistrust, and could escalate into a destructive arms race or even a war in space. It is important to remember that we would all be losers in such a case, no matter which country we are from. Space is a global resource where careless behaviour by one actor will affect all other users, including the perpetrator. Irresponsible and unilateral action in space is counterproductive for all space powers as it creates insecurity in space, fosters further unilateral defensive measures, threatens the utilization of space for future generations and ultimately affects global security on Earth.

> Because space warfare can be more indiscriminate than terrestrial warfare, and because all space-faring nations are increasingly dependent on space assets for national and economic security, all major powers face the same fundamental dilemma that satellites are both essential and extraordinarily vulnerable, and that the use of weapons in space is likely to have unintended, negative consequences. Mechanical objects may be the initial victims of space warfare, but satellites are unlikely to be the only victims, since they are directly linked to soldiers, noncombatants, and nuclear weapons.[170]

Acceptance of this mutual vulnerability in space should deter further weaponization of outer space and maximize safety and freedom of action in space.[171] The sustainability and growth of any space power will be undermined by the use of force and it is therefore in the self-interest of all space-faring nations to secure space and preserve our common heritage by addressing all five dimensions of global security.[172]

As noted earlier in my definition of space power, outer space will either be safe for every one or for no one.

'Symbiotic realism': Relevance to space governance

Given increasing globalization, interdependence and interconnectivity, it becomes clear that one actor cannot create a lasting and secure environment in space on its own and requires a multilateral approach. One such approach, advocated previously as a *'symbiotic realism'* theory of international relations might be useful.[173] This approach is built on the assumption that the international system is the result of four interlocking dimensions. These are (a) the neurobiological substrates of human nature, (b) the global state of nature (global anarchy), (c) interdependence and (d) instant connectivity that results from globalization. I defined the neurobiological substrates of human nature that motivate behaviour as basic needs, ego and fear. When basic survival needs are met, I argue that humans can aspire to higher things such as morality. Thus, in order for society to prosper, the state of nature among individuals must be mitigated. This has historically been done through the establishment of states and of domestic governments.

What makes this *'symbiotic realism'* framework different from other realist or neoliberal approaches is that it assumes that state behaviour is driven by 'emotionality and perceptual frameworks' as well as by 'rationality'. *'Symbiotic realism'* accounts for the role of inter-subjective meanings in the constitution of material power or capabilities. It also captures how part of a state's power may come from its capacity to generate consent, which involves cognitive and normative schemes as well as material phenomena. It indicates that hegemony (on Earth and in outer space), based on consent and symbiotic relations may provide the best means, at present, of mitigating the consequences of the interlocking dimensions of states' emotionality, globalization, insecurity and global anarchy. It advocates a symbiotic interdependence between states in a connected and interdependent world, whose security has become intertwined, especially in a global commons domain like outer space. It does this while allowing some states to gain more than others through non-conflictual competition and without deleterious effects.

In other words, *'symbiotic realism'* addresses the rational interests of states, like power relations and national security imperatives, as well as the emotional and perceptual frameworks aspects of states' behaviour, like pride, prestige, fear, insecurity, cultural and national arrogance and exceptionalism, and other cyclical domestic populist pressures, which have always been strong motivators of politics and policy in general and of space politics and policy in particular.

Internationally, the relations between states have been historically and continue to be dominated by anarchy. With no overarching authority to regulate state behaviour and ensure the safety and prosperity of all, international life could be considered somewhat precarious. I argue that increased integration brought about by globalization can help mitigate the consequences of global anarchy. However, globalization is also undermining the capacities of states to act as viable sites for collective action and credible commitments. This is because states are becoming more intertwined in webs of power that are linked to shifts in the material distribution of power and authoritative resources. In other words, the state's domestic role is being transformed by the increasingly important interests of transnational capital. Because of these new emerging dynamics of the international system, I maintain that in order for societies to prosper materially and morally, humanity needs some form of governance that can ensure that the basic survival needs of human beings and states are satisfied in the context of increased interdependence and instant connectivity. This conception of international relations allows for absolute rather than just relative gains, and stresses that it is in the interests of all actors to cooperate with rather than oppose each other.

The '*symbiotic realism*' theory of international relations goes beyond the state-centrism of realism, integrating a number of actors that have often been underemphasized or ignored by the realist paradigm, such as large collective identities, multinational corporations, international organizations, the biosphere and women. This is vital, since these, too, are all important actors (both active and reactive) that help to (re)produce the global order and, as such, have a bearing on its relations and dynamics.

Without international regulations, the anarchic organization of the space realm, combined with the amorality of state behaviour due to the presence of survival instincts and the prioritization of national interests make continued competition in space inevitable. Yet, in line with the '*symbiotic realism*' theory, as long as this competition remains non-conflictual, it does not represent any threat to the security and sustainability of space. On the contrary, non-conflictual competition benefits space actors as it fosters technological innovation and progress, and allows the development of new applications at lower costs, thus making space technology accessible to a larger audience. However, given the innate national interests of national powers and their persistent focus on national security, it is highly likely that without robust, comprehensive space treaties and international regulations to govern the use of space, emerging and established space powers will launch more ASAT tests and other provocative measures, undermining space security and everyday uses of space. These grim predictions have been confirmed by the outcomes of the *meta*-geopolitics analysis in Chapter 5. The comparative analysis of existing geopolitical configurations in space showed that emerging space powers strive to demonstrate their newly achieved space capabilities while established space players seek to sustain and further enhance their own capabilities. Security threats against space assets and future activities in space are therefore serious and if no action is taken, such critically strategic assets as GPS, communications and remote sensing satellites, mobile telephones and the Internet will remain vulnerable.

Policy recommendations

Throughout this book, it has become obvious that we are more and more dependent on space for providing security while at the same time we face increasingly serious challenges to space assets. It is important to build a comprehensive framework to tackle these challenges, where justice and good governance guide the common efforts of all states.[174]

Cooperation efforts and the creation of a comprehensive regulatory framework for space activities could take place in existing international forums, such as COPUOS and the CD. However, the existing international negotiating bodies are hampered by conflicting national interests in security and defence matters. The close relationships between space and security, and defence and military capabilities, make any space-related negotiations a minefield of complicated and controversial topics. If such increased cooperation efforts and possible negotiations could be initiated through confidence-building measures rather than a comprehensive treaty, issues such as space debris, launch notifications and sustainable increases in space traffic could set a precedent for future commitments and increase the common interest in and benefits of space.

In addition to threatening space events such as increased debris or space weaponization, there is also the dual-use dilemma of peaceful space technologies. Space launch programmes are being developed in countries where long-range ballistic missile capacities could rapidly change geopolitical situations as well as regional and global security. Space has been seen as a separate sphere, delinked from other types of activities. Today, however, space applications are much more integrated with activities on Earth, and used to develop our societies. Space policy and space security must enter a new phase in which they are integrated into an overall security enhancement effort. It is essential that space institutions, both national and international, reinforce their communication channels with each other. At the same time, however, it is important to take into account that space technologies provide excellent opportunities for enhancing security. Such technologies provide faster information, communications, early warning, environmental assessments, intelligence gathering, monitoring capabilities and many other improvements that can be used on Earth to enhance global stability and security.

The proposed '*good governance model*' of space, which is a combination of the '*symbiotic realism*' theory of international relations and the '*multi-sum security principle*', allows a symbiotic non-conflictual competition in space. This type of competition strengthens the vitality of space and reduces the cost of access to space and space technology. It enhances states' capacities in numerous domains, such as social and economic development, climate change monitoring, disaster management and international treaty verification. Absolute gains and non-conflictual competition also foster scientific research and technological progress, which enhance our understanding of our planet and the solar system, and help global leaders improve the quality of life of their citizens and resolve important issues on Earth. Limited resources on Earth have made sustainability a key issue for policymakers. Space applications can provide assistance with such challenges and problems but in order to further explore the opportunities that space gives us, its potential and its inherent value for all of mankind need to be realized.

In order to create such a space governance model, the recommendations below may be helpful to policymakers, academics, NGOs, scientists, international organizations and other groups interested in outer space:

1. Given the numerous obstacles to the emergence of an overarching legally binding space treaty and the urgency of addressing security issues in space, the development of soft law initiatives, such as codes of conduct and any other efforts aimed at enhancing TCBMs, would lay the basis for future, more robust measures. Such voluntary rules could help prevent irresponsible behaviour and facilitate action being taken against the perpetrators of harmful activities in space. It is critical that all major space powers, including the United States,

the EU, Russia and China support such initiatives and strive to develop a common proposal.

2. It is critical that future international discussions on space security address the issue of ground-based ASATs, including the testing, development and deployment of such measures. Other asymmetrical techniques such as lasers, jamming and cyber threat as well as the dual-use dilemma of space-related assets also require increased attention and enhanced regulation.
3. There is also a critical need to address the legal implications of the increased commercial space activity for issues such as property, liability, definitions of launching state and insurance. The development of commercial space tourism and private manned space flights need to be taken into consideration.
4. There is a pressing need to address the space traffic issue. US technological superiority in space allows it to provide leadership in the traffic management of space objects. This applies to satellites as well as debris. The ability of the United States to track space objects has in the past provided very useful warning to avoid collisions. This space traffic management role is vital for the security of space objects as well as the overall global governance agenda of space. This will need to include critical issues such as (a) debris mitigation; (b) collision avoidance; (c) spectrum allocation and (d) space situational awareness issues. This will need a collective collaborative transnational effort built on transparency, real-time information exchange, consultation and verification mechanisms and continued technological advances in monitoring capabilities.
5. Increased resources and support for civil society and scientific community initiatives would produce valuable and innovative technological solutions, for example, for removing or de-orbiting critical pieces of debris. Shared international constellations maximize the financial and strategic benefits of emerging small space capabilities such as micro-satellites. Such cooperation is particularly important in enabling developing countries to acquire affordable space capabilities. Furthermore, the multi-stakeholder ventures create mutual dependencies and help deter threatening behaviours.
6. Increased coordination and closer cooperation between the multilateral bodies dealing with space matters, COPUOS, the ITU and the CD are necessary to avoid duplication and cross purposes. It will remove the artificial barriers used by states to separate their military, civilian and commercial space operations and provide a holistic approach to space issues.

7. Space policy and space exploration must be guided by the eight principles of good global governance that I advocated previously, such as (a) effective multilateralism, (b) effective multilateral institutions, (c) representative multilateral decision-making structures, (d) dialogue, (e) accountability, (f) transparency, (g) burden-sharing and (h) stronger partnerships between multilateral organizations and civil society.[175]

Conclusions

As one of the most fundamental issues of the twenty-first century, space is and will continue to be a key driver of technological developments and enhanced capabilities. The challenges are many and the possible solutions to the dilemmas posed by actions and inaction in space are limited. My guess is that despite all the risks involved in the weaponization of space, the three major space powers (US, Russia and China) will eventually do so for the following reasons:

- global anarchy and insecurity;
- lack of a just overarching global authority and legal framework;
- national pride and prestige, which I call the *emotionality of states*; and finally
- the pursuit of national interests, which I call the *egoism of states*.

Based on these motivations, the likelihood of the weaponization of space is high and I think inevitable. The United States is the most likely candidate to do it first because of its current capabilities, resources and need to remain dominant (linked to both pride and national security imperatives). Russia and China will then do the same. The consequences of this inevitable weaponization of space will be an insecure world and a vulnerable and uncertain future for humanity.

In the face of fast developing space technologies and countries' increasing reliance on satellite technology, a robust legal instrument aimed at preventing a space arms race is urgently needed. As I noted in a previous book, '[s]hould the international community fail to agree on such a treaty or fail to get all technologically sophisticated military powers on board, the militarization of space is almost inevitable. This would radically alter the geopolitics as we know it, for whoever will dominate space will come to dominate the Earth.'[176]

Given that human behaviour is primarily governed by *'emotional amoral egoism'*,[177] it is unrealistic to expect cooperative behaviour even if this appears to be the logical or rational thing to do. At the state level and in the context of outer space power dynamics, *state emotionality* is displayed as national pride and *state egoism* represents the national interest. The *emotional egoism* of states will be pursued within an 'amoral framework' (if not worse) and may even sabotage the very national interest that it is supposed to serve in the long run, due to short-term electoral domestic pressure. In view of this, our best hope for a peaceful outer space environment is not to expect cooperative behaviour from states (even this appears to be in their interests) but to aim for absolute gains and a non-conflictual competitive environment, governed by soft law rules and empowered by the '*symbiotic realism*' framework of international relations and the '*multi-sum security principle*' of global security, discussed above.

The most significant step is to truly consider space as a global common. There are similar notions in place already, such as the significant increase in international cooperation and in the number of multinational companies providing global services. These events create shared benefits and common interests and efficiently promote multilateral responses to challenges and threats. International cooperation initiatives also have the benefit of increasing trust, transparency and confidence between parties. This could discourage the view that unilateral action is an appropriate way to provide security.

The most important challenge for governments, businesses and society is to strike a balance between the need to regulate and oversee individual behaviours in space (especially the potentially dangerous ones) while still encouraging non-conflictual competition that is essential for driving progress, innovation and decreasing the cost of space access and its utilization. In order to regulate security-threatening behaviours, arms control treaties, codes of conduct, political measures such as extended communication between parties, cooperation through organizations and confidence-building measures must be pursued, taking into account future developments and dual-use possibilities. Multilateral institutions and rules are needed in order to deter individual states from resisting collective action and making dangerous unilateral moves.

Notes

1 Space Power: An Introduction

1. N. R. F. Al-Rodhan (2009), *Neo-Statecraft and Meta-Geopolitics: Reconciliation of Power, Interests, and Justice in the 21st Century* (Berlin: LIT), p. 51.
2. G. L. Schulte (2011), 'National Space Security Strategy Outlines Rules of the Road', American Forces Press Service, 11 February, http://www.defense.gov//News/NewsArticle.aspx?ID=62791, date accessed 26 July 2011.
3. UNIDIR (2010), 'Space Security 2010: From Foundations to Negotiations', Conference Report 29–30 March (New York and Geneva: United Nations), http://www.unidir.org/pdf/ouvrages/pdf-1-92-9045-010-C-en.pdf, date accessed 26 July 2011, p. 12.
4. Ibid., p. 15.
5. J. Moore (2011), 'Crowded Cosmos Calls for National Security Space Strategy', *Executive Gov*, 8 April, http://www.executivegov.com/2011/04/crowded-cosmos-calls-for-national-security-space-strategy/, date accessed 26 July 2011.
6. Cf. Al-Rodhan (2009), *Neo-Statecraft and Meta-Geopolitics.*
7. Ibid., p. 33.
8. M. Heffernan (2000), 'Fin du siècle, fin du monde?', in K. Dodds and D. Atkinson (eds), *Geopolitical Traditions: A Century of Geopolitical Thought* (London: Routledge), p. 28.
9. J. J. Grygiel (2006), *Great Powers and Geopolitical Change* (Baltimore: John Hopkins University Press), p. 5.
10. Heffernan (2000), 'Fin du siècle, fin du monde?' in Dodds and Atkinson (eds) (2000), *Geopolitical traditions,* p. 28.
11. J. Havercroft and R. Duvall (2009), 'Critical Astropolitics: The Geopolitics of Space Control and the Transformation of State Sovereignty', in N. Bormann and M. Sheehan (eds) *Securing Outer Space* (New York and Milton Park: Routledge), p. 43.
12. Grygiel (2006), *Great Powers and Geopolitical Change,* p. 7.
13. Ibid., p. 9.
14. 'Q&A: Sputnik' (2007), BBC News, 3 October, http://news.bbc.co.uk/2/hi/6937964.stm, date accessed 7 February 2011.
15. J. C. Moltz (2008), *The Politics of Space Security: Strategic Restraint and the Pursuit of National Interests* (Stanford: Stanford University Press), p. 23.
16. M. Sheehan (2007), *The International Politics of Space* (New York and London: Routledge), p. 7.
17. J. A. Vasquez (1998), *The Power of Power Politics: From Classical Realism to Neotraditionalism* (Cambridge: Cambridge University Press), p. 13.
18. H. J. Morgenthau (1960), *Politics among Nations: The Struggle for Power and Peace* (New York: Alfred A. Knopf), p. 3–15.
19. Sheehan (2007), *The International Politics of Space,* p. 8.

20. N. R. F. Al-Rodhan (2007), *Symbiotic Realism: A Theory of International Relations in an Instant and an Interdependent World* (Berlin: LIT), p. 22.
21. R. L. Pfaltzgraff, Jr. (2011), 'International Relations Theory and Spacepower', in C. D. Lutes, P. L. Hays (eds), with V. A. Manzo, L. M. Yambric and M. E. Bunn, *Toward a Theory of Spacepower: Selected Essays* (Washington, D.C.: National Defense University Press), http://www.ndu.edu/press/spacepower.html, date accessed 6 June 2011, p. 36.
22. Ibid.
23. Cf. K. N. Waltz (1979), *Theory of International Politics* (New York: McGraw-Hill).
24. E. C. Dolman (2002), *Astropolitik: Classical Geopolitics in the Space Age* (Portland and London: Frank Cass Publishers), p. 2.
25. Ibid., p. 15.
26. Ibid.
27. Moltz (2008), *The Politics of Space Security,* pp. 25–6.
28. Dolman (2002), *Astropolitik,* p. 2.
29. Ibid., p. 8.
30. Moltz (2008), *The Politics of Space Security*, p. 25.
31. Ibid.
32. Ibid., pp. 25–6.
33. Dolman (2002), *Astropolitik*, p. 15.
34. Ibid., p. 157.
35. Ibid., pp. 111–44.
36. Ibid., p. 165.
37. Ibid.
38. Ibid.
39. Ibid., p. 183.
40. Havercroft and Duvall (2009), 'Critical Astropolitics: The Geopolitics of Space Control and the Transformation of State Sovereignty', in Bormann and Sheehan (2009) *Securing Outer Space*, p. 51.
41. Al-Rodhan (2007), *Symbiotic Realism,* p. 25.
42. Sheehan (2007), *The International Politics of Space,* p. 12.
43. Al-Rodhan (2007), *Symbiotic Realism*, p. 28.
44. Ibid., p. 32.
45. Sheehan (2007), *The International Politics of Space,* p. 15.
46. Ibid., p. 12.
47. Cf. D. Deudney (1983), 'Whole Earth Security: A Geopolitics of Peace', *World Watch Paper 55* (Washington, D.C.: Worldwatch Institute).
48. Havercroft and Duvall (2009), 'Critical Astropolitics: The Geopolitics of Space Control and the Transformation of State Sovereignty', in Bormann and Sheehan (2009) *Securing Outer Space*, p. 48.
49. Ibid.
50. Ibid.
51. Ibid.
52. Ibid.
53. Ibid., p. 47.
54. Moltz (2008), *The Politics of Space Security*, p. 28.
55. Ibid., p. 30.

56. Ibid.
57. Ibid.
58. Sheehan (2007), *The International Politics of Space*, p. 13.
59. Moltz (2008), *The Politics of Space Security*, p. 29.
60. Ibid., p. 31.
61. Ibid., p. 32.
62. For a more detailed application of economically based convergence theory, see Harold L. Wilensky (2002), *Rich Democracies: Political Economy, Public Policy, and Performance* (Berkeley: University of California Press). The same technological view was used by Soviet space analysts with the final aim of the emergence of a harmonious communist society. For more details of the Soviet school, see V. S. Vereshchetin (1977), *Mezdunarodnoe sotrudnichestvo v kosmose* [International Cooperation in Space] (Moscow: Nauka); P. I. Lukin et al. (1980), *Kosmos I pravo* [Space and law] (Moscow: Institute of State and Law) and G. P. Zhukov (1985), *Kosmos I mir* [Space and peace] (Moscow: Nauka). A more pessimistic school of technological determinism emerged in the 1950s. For more details, see A. W. Frutkin (1965), *International Cooperation in Space* (Englewood Cliffs NJ: Prentice-Hall); H. F. York (1970), *Making Weapons, Talking Peace: A Physicist's Odyssey from Hiroshima to Geneva* (New York: Basic).
63. Moltz (2008), *The Politics of Space Security*, p. 32.
64. Per Magnus Wijkman (1982), 'Managing the Global Commons', *International Organization*, vol. 36, issue 3, pp. 511–36.
65. Ibid., p. 535.
66. Moltz (2008), *The Politics of Space Security*, pp. 36–7.
67. Cf. P. B. Stares (1987), *Space and National Security* (Washington, D.C.: Brookings Institution).
68. Moltz (2008), *The Politics of Space Security*, p. 38.
69. Cf. M. Krepon and C. Clary (2003), *Space Assurance or Space Dominance? The Case Against Weaponizing Space* (Washington, D.C.: Henry L. Stimson Center).
70. Sheehan (2007), *The International Politics of Space*, p. 15.
71. Ibid., p. 12.
72. Ibid., p. 13.
73. Ibid., p. 16.
74. Ibid., p. 13.
75. Havercroft and Duvall (2009), 'Critical Astropolitics', in Bormann and Sheehan (2009) *Securing Outer Space*, p. 50.
76. Ibid.
77. Cf. G. Ó Tuathail (1996), *Critical Geopolitics: The Politics of Writing Global Space* (Minneapolis: University of Minnesota Press).
78. Al-Rodhan (2007), *Symbiotic Realism*, p. 36.
79. M. Kuus (2009) 'Critical Geopolitics', *The International Studies Association Compendium Project*, http://www.heinebuch.de/include/Thema1322/pdf/02.pdf, date accessed 10 August 2011, p. 3.
80. Sheehan (2007), *The International Politics of Space*, p. 17.
81. Havercroft and Duvall (2009), 'Critical Astropolitics', in Bormann and Sheehan (2009) *Securing Outer Space*, p. 51.

82. Ibid., p. 56.
83. Ibid., p. 57.
84. D. Grondin (2009), 'The (Power) Politics of Space: The US Astropolitical Discourse of Global Dominance in the War of Terror', in Bormann and Sheehan (2009) *Securing Outer Space*, p. 111.
85. Al-Rodhan (2009), *Neo-Statecraft and Meta-Geopolitics*, p. 51.
86. Ibid, p. 153, pp. 155–6.

2 Space Power and *Meta*-Geopolitics

1. N. R. F. Al-Rodhan (2009), *Neo-Statecraft and Meta-Geopolitics. Reconciliation of Power, Interests and Justice in the 21st Century* (Berlin: LIT), p. 202.
2. Ibid.
3. Ibid., p. 53.
4. Cf. Al-Rodhan (2009), *Neo-Statecraft and Meta-Geopolitics.*
5. Ibid., p. 51.
6. Ibid., p. 79.
7. 'Q&A: Sputnik' (2007), BBC News, 3 October, http://news.bbc.co.uk/2/hi/6937964.stm, date accessed 7 February 2011.
8. C. Clover (1999), 'Dreams of the Eurasian Heartland: The Reemergence of Geopolitics', *Foreign Affairs,* March/April, http://www.foreignaffairs.com/articles/54793/charles-clover/dreams-of-the-eurasian-heartland-the-reemergence-of-geopolitics, date accessed 6 January 2011.
9. M. Sheehan (2007), *The International Politics of Space* (Oxford: Routledge), p. 113.
10. D. R. Baucom (1992), *Clausewitz on Space War: An Essay on the Strategic Aspects of Military Operations in Space* (Maxwell Air Force Base, Alabama: Air University Press), p. 3.
11. J. J. Jusell (1998), *Space Power Theory: A Rising Star,* Research Report (Maxwell Air Force Base, Alabama: Air University Press), http://www.fas.org/spp/eprint/98-144.pdf, date accessed 17 December 2010, p. 10.
12. E. Bergaust (1964), *The Next Fifty Years in Space* (New York: Macmillan & Co.), cited in Jusell (1998), *Space Power Theory*, p. 10.
13. M. E. B. France (2000), 'Back to the Future: Space Power Theory and A. T. Mahan', *Space Policy*, November, vol. 16, issue 4, p. 237.
14. Cf. A. T. Mahan (1918), *The Influence of Sea Power upon History, 1660–1783* (Boston: Little, Brown).
15. J. Oberg (1998), *Space Power Theory* (Washington, D. C.: Government Printing Office), p. 121.
16. Ibid.
17. Jusell (1998), *Space Power Theory,* p. 12.
18. F. J. Gayl (2004), 'Time for a Military Space Service', *Proceedings*, vol. 130, issue 7, July, p. 44.
19. Oberg (1998), *Space Power Theory*, p. 122.
20. Ibid., p. 121.
21. Ibid.
22. Ibid., p. 6.

23. Secretary of the Air Force (1998), *Space Operations: Air Force Doctrine Document 2-2*, HQ Air Force Doctrine Command, United States, 23 August, http://www.fas.org/man/dod-101/usaf/docs/afdd/afdd2-2.pdf, date accessed 27 July 2011, p. 1.
24. D. J. Johnson, S. Pace and C. C. Gabbard (1998), *Space: Emerging Options for National Power* (Santa Monica: Rand), p. 8.
25. J. M. Logsdon (2001), 'Just Say Wait to Space Power', *Issues in Science and Technology*, vol. 17, issue 3, http://www.issues.org/17.3/p_logsdon.htm, date accessed 7 January 2011.
26. Oberg (1998), *Space Power Theory*, p. 10.
27. J. E. Shaw (1999), 'The Influence of Space Power upon History (1944–1998)', *Air Power History*, vol. 46, issue 4, Winter, pp. 20–9.
28. J. Nye (2004), *Soft Power: The Means to Success in World Politics* (New York: Public Affairs), p. 8.
29. France (2000), 'Back to the Future', p. 239.
30. Ibid.
31. Ibid.
32. Ibid., p. 240.
33. Oberg (1998), *Space Power Theory*, pp. 44–6.
34. Ibid., pp. 45–7.
35. N. Peter (2009), 'Space Power and Europe in the 21st Century', *European Space Policy Institute Perspectives* 21 April, http://www.espi.or.at/images/stories/dokumente/Perspectives/ESPI_Perspectives_21.pdf, date accessed 27 July 2011, p. 2.
36. R. L. Pfaltzgraff, Jr. (2011), 'International Relations Theory and Spacepower' in C. D. Lutes, P. L. Hays (eds), with V. A. Manzo, L. M. Yambric and M. E. Bunn, *Toward a Theory of Spacepower: Selected Essays* (Washington, D.C.: National Defense University Press), http://www.ndu.edu/press/spacepower.html, date accessed 6 June 2011, p. 43.
37. N. Peter (2008), 'Space Power and Europe: In the Need for a Conceptual Framework', *Proceedings of the IAC, International Astronautical Congress*, http://www.espi.or.at/images/stories/dokumente/presentations/2008/iac-08-e3.2.9-nicolas_peter.pdf, date accessed 27 July 2011, p. 5.
38. Pfaltzgraff (2011), 'International Relations Theory and Spacepower', in Lutes and Hays (eds), et al. (2011), *Toward a Theory of Spacepower*, p. 29.
39. Al-Rodhan (2009), *Neo-Statecraft and Meta-Geopolitics*, p. 60.
40. P. K. Misra (2010), 'A Dedicated Satellite for Meeting Health Education Needs of Afro-Asian Nations: Possibilities, Action Plans and Benefits', *Electronic Journal of Information Systems in Developing Countries* (EJISDC), vol. 41, issue 6, http://www.ejisdc.org/ojs2/index.php/ejisdc/article/viewFile/630/319, date accessed 27 July 2011, pp. 1–12.
41. Ibid.
42. B. Rao and A. Lombardi (2009), 'Telemedicine: Current Status in Developed and Developing Countries', *Journal of Drugs in Dermatology*, April, http://www.thefreelibrary.com/_/print/PrintArticle.aspx?id=201673296, date accessed 6 December 2010.
43. Cf. Misra (2010), 'A Dedicated Satellite for Meeting Health Education Needs of Afro-Asian Nations'.

44. 'Remote Diagnosis' (2009), *The Economist*, 21 May, http://www.economist.com/node/13688152?story_id=13688152, date accessed 25 November 2010.
45. Cf. Al-Rodhan (2009), *Neo-statecraft and Meta-geopolitics.*
46. P. J. Garrity (1998), 'Interests and Issues: Perspectives in Future Challenges to US Security', in S. A. Cambone (ed) *A New Structure for National Security Policy Planning* (Washington, D.C.: Center for Strategic and International Studies), p. 101.
47. Space Foundation (2011), 'The Space Report 2011 Reveals Continued Space Sector Growth Driven by Commercial Businesses', *Media*, 7 April, http://www.spacefoundation.org/news/story.php?id=1100, date accessed 21 July 2011.
48. R. C. Barbosa (2010), 'China Opens 2010 with BeiDou-2 Satellite Launch', *Nasa Spaceflight.com*, 16 January, http://www.nasaspaceflight.com/2010/01/china-opens-2010-with-beidou-2-satellite-launch, date accessed 10 August 2011.
49. W. J. Broad and K. Chang (2010), 'Obama Reverses Bush's Space Policy', *New York Times*, 28 June, http://www.nytimes.com/2010/06/29/science/space/29orbit.html, date accessed 28 July 2011.
50. N. R. F. Al-Rodhan (2009), 'Balancing Transnational Responsibilities and Burden-sharing with Sovereignty and Human Dignity', in N. R. F. Al-Rodhan (ed), *Potential Global Strategic Catastrophes: Balancing Transnational Responsibilities and Burden-sharing with Sovereignty and Human Dignity* (Berlin: LIT), pp. 21–30.
51. W. H. Lambright (2007), 'NASA and the Environment: Science in a Political Context', in S. J. Dick and R. D. Launius (eds), *Societal Impacts of Space Flight* (Washington, D.C.: NASA), http://history.nasa.gov/sp4801-chapter16.pdf, date accessed 12 December 2010, p. 314.
52. Ibid., p. 328.
53. M. Sonter (2006), 'Asteroid Mining: Key to the Space Economy', *Space.com*, 9 February, http://www.space.com/adastra/060209_adastra_mining.html, date accessed 12 December 2010.
54. L. I. Tennen (2010), 'Towards a New Regime for Exploitation of Outer Space Mineral Resources', *Nebraska Law Review*, vol. 88, issue 794, p. 794.
55. Al-Rodhan (2009), *Neo-Statecraft and Meta-Geopolitics,* p. 70.
56. G. Acar (2009), 'The Geopolitics of Satellite Navigation Technology: The Russian alternative to GPS', *Eurasia Critic*, March, http://www.eurasiacritic.com/articles/geopolitics-satellite-navigation-technology-russian-alternative-gps, date accessed 10 December 2010.
57. A. M. Saperstein (2002), '"Weaponization" vs. "Militarization" of Space', *US Physical Society, Forum on Physics & Society*, vol. 31, issue 3, July, http://aps.org/units/fps/newsletters/2002/july/saperstein.pdf, date accessed 11 December 2010; C. Peoples (2008) 'Assuming the Inevitable? Overcoming the Inevitability of Outer Space Weaponization and Conflict', *Contemporary Security Policy*, vol. 29, issue 3, http://tandfprod.literatumonline.com/doi/pdf/10.1080/13523260802514811, date accessed 27 July 2011, p. 502.
58. B. D. Watts (2001), *The Military Use of Space: A Diagnostic Assessment* (Washington, D.C.: Center for Strategic and Budgetary Assessments), http://www.csbaonline.org/wp-content/uploads/2011/02/2001.02.01-Military-Use-of-Space.pdf, date accessed 27 July 2011, p. 1.

59. M. Krepon, T. Hitchens and M. Katz-Hyman (2011), 'Preserving Freedom of Action in Space: Realizing the Potential and Limits of US Spacepower', in Lutes and Hays (eds), et al. (2011), *Toward a Theory of Spacepower*, p. 395.
60. Saperstein (2002), '"Weaponization" vs. "Militarization" of Space', p. 1.
61. N. R. F. Al-Rodhan (2007), *The Five Dimensions of Global Security: Proposal for a Multi-sum Security Principle* (Berlin: LIT), p. 137.
62. Ibid., pp. 84–5.
63. Ibid., p. 131.
64. E. Bahr (1985), 'Observations on the Principle of Common Security', in R. Vayrynen (ed), *Policies for Common Security* (London: Taylor & Francis), p. 31.
65. J. K. Hettling (2003), 'The Use of Remote Sensing Satellites for Verification in International Law', *Space Policy*, vol. 19, issue 1, p. 33.
66. I. Niemeyer (2009), 'Perspectives of Satellite Imagery Analysis for Verifying the Nuclear Non-Proliferation Treaty', in B. Jasani, I. Niemeyer, S. Nussbaum (eds), *International Safeguards and Satellite Imagery* (Berlin: Springer), p. 35.
67. Cf. United Nations Framework Convention on Climate Change (UNFCC), 'Kyoto Protocol', http://unfccc.int/kyoto_protocol/items/2830.php, date accessed 12 December 2010.
68. Cf. UNFCC, 'Systematic Observation of the Kyoto Protocol', http://unfccc.int/methods_and_science/research_and_systematic_observation/items/3462.php, date accessed 12 December 2010.
69. Cf. N. R. F. Al-Rodhan (2007), *Symbiotic Realism: A Theory of International Relations in an Instant and an Interdependent World* (Berlin: LIT).
70. Al-Rodhan (2007), *The Five Dimensions of Global Security*, p. 133.
71. Ibid.

3 Space Technology and *Meta*-Geopolitics

1. B. De Montluc (2009), 'The New International Political and Strategic Context for Space Policies', *Space Policy*, vol. 25, issue 1.
2. N. Peter (2009), 'Space Power and Europe in the 21st Century', *European Space Policy Institute Perspectives*, issue 21, April, http://www.espi.or.at/images/stories/dokumente/Perspectives/ESPI_Perspectives_21.pdf, date accessed 27 July 2011, p. 2.
3. A. Lebeau (2008), 'Space: The Routes of the Future', *Space Policy*, vol. 24, issue 1, February, p. 43.
4. M. Sheehan (2007), *The International Politics of Space* (New York and London: Routledge), pp. 20–1.
5. D. D. Stine (2007), *US Civilian Space Policy Priorities: Reflections 50 Years after Sputnik. CRS Report for Congress*, December, http://books.google.ch/books?id=U2IAyxWVF0wC&printsec=frontcover&hl=en&source=gbs_ge_summary_r&cad=0#v=onepage&q&f=false, date accessed 2 August 2011, p. 2.
6. Cf. D. S. F. Portree (1998), 'NASA's Origins and the Dawn of the Space Age', *NASA Monographs in Aerospace History*, issue 10, http://www.hq.nasa.gov/office/pao/History/monograph10/, date accessed 20 December 2010.

7. Cf. W. D. Compton (1989), Where No Man Has Gone Before: A History of Apollo Lunar Exploration Missions, NASA Special Publication-4214, NASA History Series, http://www.solarviews.com/history/SP-4214/contents.html, date accessed 21 December 2010.
8. T. Hitchens (2003), *Developments in Military Space: Movement Towards Space Weapons?,* Centre for Defense Information, October, http://www.cdi.org/pdfs/space-weapons.pdf, date accessed 21 December 2010, p. 1.
9. Cf. F. G. Klotz (1998), Space, Commerce and National Security (New York: Council on Foreign Relations).
10. J. E. Shawn (1999), 'The Influence of Space Power upon History (1944–1998)', *Air Power History*, vol. 46, issue 4, Winter, pp. 20–9.
11. Cf. Klotz (1998), 'Space, Commerce, and National Security'.
12. E. S. Waldrop (2003), 'Integration of Military and Civilian Space Assets: Legal and National Security Implications', unpublished LL.M. Dissertation, McGill University, http://digitool.library.mcgill.ca/webclient/StreamGate?folder_id=0&dvs=1312278427011~816, date accessed 2 August 2011, p. 5.
13. Ibid.
14. Cf. Shawn (1999), 'The Influence of Space Power upon History (1944–1998)'.
15. Ibid.
16. Attributed to Kutny, in D. A. Day (1996), 'The Air Force in Space: Past. Present', *Space Times*, March/April, p. 18, cited in Shawn (1999) 'The Influence of Space Power upon History (1944–1998)'.
17. United Nations Security Council (1990), *Resolution 660*, S/RES/660, 2 August, http://www.fas.org/news/un/iraq/sres/sres0660.htm, date accessed 2 August 2011; United Nations Security Council (1990), *Resolution 678*, S/RES/678, 29 November, http://www.fas.org/news/un/iraq/sres/sres0678.htm, date accessed 2 August 2011.
18. Sheehan (2007), *The International Politics of Space*, pp. 99–100.
19. British Parliamentary Office of Science and Technology (2006), 'Military Uses of Space', Postnote 273, December, http://www.parliament.uk/documents/post/postpn273.pdf, date accessed 2 August 2011, pp. 1–4.
20. C. Conetta (2005) 'Arms Control in an Age of Strategic and Military Revolution', *Project on Defense Alternatives Research Monograph*, issue 11, http://www.comw.org/pda/fulltext/0511rm11.pdf, date accessed 2 August 2011, p. 5.
21. E. Sadeh (2010), 'Towards a National Space Strategy', *Astropolitics*, vol. 8, issue 2–3, p. 82.
22. M. S. Smith (2004), 'US Space Programs: Civilian, Military and Commercial', CRS Issue Brief for Congress, http://www.fas.org/spp/civil/crs/IB92011.pdf, date accessed 8 January 2011, pp. 1–17.
23. Lebeau (2008), 'Space: The Routes of the Future', p. 43.
24. J. Johnson-Freese quoted in L. David (2006), 'US-China Space Ties Weighed', *Space.com,* 20 April, http://www.space.com/news/060420_china_visit.html, date accessed 7 January 2011.
25. P. Meyer (2006), 'Opening Remarks', Report of the United Nations Institute for Disarmament Research Conference (UNIDIR), *Building the Architecture*

for Sustainable Space Security, 30–1 March (Geneva: United Nations Press), p. xxii.

26. W. von Kries, 'Dual Use of Satellite Remote Sensing', *Space4peace.org*, http://www.space4peace.org/ethics/dusrs.htm, date accessed 7 January 2011.
27. RAND National Defense Research Institute (2006), *National Security Space Launch Report* (Santa Monica: RAND Corporation), http://www.rand.org/pubs/monographs/2006/RAND_MG503.pdf, date accessed 8 January 2011, pp. 2–3.
28. Ibid.
29. R. Skaar (2007), 'Commercialisation of Space and its Evolution: Will New Ways to Share Risks and Benefits Open Up a Much Larger Space Market?', *European Space Policy Institute Report* 4, May, http://www.espi.or.at/images/stories/dokumente/studies/commercialisation.pdf, date accessed 8 January 2011, p. 9.
30. J. A. Giacalone (2008), 'Global Trends in the Commercialization of Space', *Journal of Business & Economics Research*, vol. 6, issue 8, http://cluteonline.com/journals/index.php/JBER/article/view/2462/2508, date accessed 2 August 2011, pp. 65–76.
31. Space Foundation (2008), *The Space Report 2008: The Authoritative Guide to Global Space Activity*, Executive Summary, http://www.contentfirst.com/past/Spacefoundation/08executivesummary.pdf, date accessed 2 August 2011, p. 8.
32. Ibid.
33. Ibid.
34. Giacalone (2008), 'Global Trends in the Commercialization of Space', p. 68.
35. Skaar (2007), 'Commercialisation of Space and Its Evolution', p. 27.
36. C. V. Peña and E. L. Hudgins (2002), 'Should the United States "Weaponize" Space? Military and Commercial Implications', *Policy Analysis*, issue 427, 18 March, http://www.cato.org/pubs/pas/pa427.pdf, date accessed 2 August 2011, p. 3.
37. M. D. Griffin (2007), 'The Space Economy', *NASA 50th Anniversary Lecture Series*, 17 September, http://www.spaceref.com/news/viewsr.html?pid=25452, date accessed 2 August 2011.
38. Sadeh (2010), 'Towards a National Space Strategy', p. 80.
39. NASA, 'What is a Satellite?', http://www.nasa.gov/audience/forstudents/k-4/stories/what-is-a-satellite-k4.html, date accessed 2 August 2011.
40. D. Wright, L. Grego and L. Gronlund (2005), *The Physics of Space Security: A Reference Manual* (Cambridge: American Academy of Arts and Sciences, http://www.amacad.org/publications/Physics_of_Space_Security.pdf, date accessed 8 January 2010, p. 16.
41. E. Sadeh (2011), 'Space and the Environment', in C. D. Lutes, P. L. Hays (eds), with V. A. Manzo, L. M. Yambric and M. E. Bunn, *Toward a Theory of Spacepower: Selected Essays* (Washington, D.C.: National Defense University Press), http://www.ndu.edu/press/spacepower.html, date accessed 6 June 2011, pp. 252–276.
42. B. Warf (2007), 'Geopolitics of the Satellite Industry', *Tijdschrift voor Economishe en Sociale Geographe*, vol. 98, issue 3, http://www2.ku.edu/~geography/Docs/

Barney%20papers/satellite%20geopolitics.pdf, date accessed 2 August 2011, p. 385.
43. A. Bhaskaranarayana, C. Varadarajan and V. S. Hedge (2009), 'Space-Based Societal Applications: Relevance in Developing Countries', *Acta Astronautica,* vol. 65, issue 9–10, November–December, p. 1479.
44. K. R. Sridhara Murthi, A. Bhaskaranarayana and H. N. Madhusudana (2010), 'New Developments in Indian Space Policies and Programmes: The Next Five Years', *Acta Astronautica*, vol. 66, issue 3–4, p. 335.
45. Bhaskaranarayana et al. (2009), 'Space-Based Societal Applications', p. 1480.
46. Murthi et al. (2010), 'New Developments in Indian Space Policies and Programmes', p. 334.
47. Ibid.
48. Indian Space Research Organization (ISRO), 'About ISRO', http://www.isro.org/scripts/Aboutus.aspx, date accessed 2 August 2011.
49. For detailed information about Indian space infrastructure, see Bhaskaranarayana et al. (2009), 'Space-Based Societal Applications', pp. 1479–85.
50. H. V. Pant and G. Bharath (2008), 'India's Emerging Missile Capability: The Science and Politics of Agni-III', *Comparative Strategy*, vol. 27, issue 4, July-September, pp. 378–9.
51. Murthi et al. (2010), 'New Developments in Indian Space Policies and Programmes', p. 334.
52. Bhaskaranarayana et al. (2009), 'Space-Based Societal Applications', p. 1481.
53. Ibid., p. 1482.
54. Sheehan (2007), *The International Politics of Space*, p. 147.
55. INSAT is one of the largest domestic communications satellite constellations in the region. It has nine operational satellites in geostationary orbit. For detailed information, see Bhaskaranarayana et al. (2009), p. 1481.
56. Bhaskaranarayana et al. (2009), 'Space-Based Societal Applications', pp. 1483–4.
57. Murthi et al. (2010), 'New Developments in Indian Space Policies and Programmes', p. 334.
58. Sheehan (2007), *The International Politics of Space*, p. 147.
59. Murthi et al. (2010), 'New Developments in Indian Space Policies and Programmes', p. 334.
60. Ibid., p. 335.
61. Bhaskaranarayana et al. (2009), 'Space-Based Societal Applications', p. 1481.
62. 'Space Programme Can Help Realise Sustainable Development: PM' (2011), *Indo-Asian News Service*, 26 March, http://www.sify.com/news/space-programme-can-help-realise-sustainable-development-pm-news-national-ld0v4idhehd.html, date accessed 7 April 2011.
63. 'Space Transportation' (2001), *US Department of Defense Space Technology Guide*, FY 2000–2001, http://space.au.af.mil/guides/stg/stg_transportation.pdf, date accessed 8 January 2011.
64. Lebeau (2008), 'Space: The Routes of the Future', p. 43.
65. 'Alcantara Launch Center (2°20'S; 44°24'W)', Global Security website, http://www.globalsecurity.org/space/world/brazil/alcantara.htm, date accessed 2 August 2011.
66. Ibid.

67. S. Pagkratis (2011), *Space Policies, Issues and Trends in 2010/2011*, ESPI Report 35(Vienna: ESPI), p. 23.
68. S. Cass (2010), 'A New Bubble for the Space Launch Industry? Iridium Boosts SpaceX, but Will the Industry Repeat the Boom-and-Bust of the 1990s?', MIT *Technology Review*, 17 June, http://www.technologyreview.com/blog/deltav/25344/, date accessed 3 January 2011.
69. 'Space Mission Generating National Pride but also Criticism' (2005), *Asian News.it*, 13 October, http://www.asianews.it/news-en/Space-mission-generating-national-pride-but-also-criticism-4350.html, date accessed 7 January 2011.
70. P. Tarikhi (2009), 'Iran's Space Programme: Riding High for Peace and Pride', *Space Policy*, vol. 25, issue 3, p. 161.
71. Ibid., p. 162.
72. Ibid., p. 163.
73. Ibid.
74. Ibid.
75. U. Rubin (2006), *The Global Reach of Iran's Ballistic Missiles*, Memorandum 86, Institute for National Security Studies (INSS), Tel Aviv University, http://www.inss.org.il/upload/(FILE)1188302022.pdf, date accessed 2 August 2011, p. 45.
76. Ibid.
77. Ibid.
78. Ibid., 45–6.
79. R. Tait (2009), 'Iran Launches First Domestically Produced Satellite', *Guardian.co.uk*, 3 February, http://www.foxnews.com/politics/2009/02/03/pentagon-irans-domestic-satellite-launch-grave-cause-concern/, date accessed 4 January 2011.
80. Tarikhi (2009), 'Iran's Space Programme', p. 163.
81. Ibid.
82. Ibid., p. 163.
83. 'Pentagon: Iran's Domestic Satellite Launch Is Grave Cause for Concern' (2009), *Foxnews*, 3 February, http://www.foxnews.com/politics/2009/02/03/pentagon-irans-domestic-satellite-launch-grave-cause-concern/, date accessed 4 January 2011.
84. Tarikhi (2009), 'Iran's Space Programme', p. 172.
85. Ibid., p. 166.
86. Ibid., p. 160.
87. Ibid., p. 173.
88. Cf. National Academy of Science (1997), 'The Global Positioning System: The Role of Atomic Clocks', *Beyond Discovery: The Path from Research to Human Benefit* (Washington, D.C.: National Academy Press), http://www.beyonddiscovery.org/content/view.page.asp?I=1275, date accessed 8 January 2011.
89. K. Stoffl (2008), 'Galileo Goes Global: What to Expect from Europe in Establishing the Navigation Satellite "System-of-Systems"', *ESPI Perspectives*, issue 9, July, http://www.espi.or.at/images/stories/dokumente/Perspectives/espi-perspectives-9.pdf, date accessed 3 August 2011, p. 1.
90. Ibid.
91. Ibid., pp. 1–2.
92. 'Galileo: A Big Opportunity for the EU's Economy' (2010), *Enterprise & Industry Online Magazine*, 16 February, http://ec.europa.eu/enterprise/

magazine/articles/industrial-policy/article_9981_en.htm, date accessed 3 August 2011.

93. 'Will China Disrupt Europe's Global Navigation Satellite System?' (2010), *GPS Systems.Net*, 10 February, http://gpssystems.net/china-disrupt-europes-global-navigation-satellite-system/, date accessed 14 August 2011.
94. B. M. DeBlois, R. L. Garwin, R. S. Kemp and J. C. Marwell (2004), 'Space Weapons: Crossing the US Rubicon', *International Security*, vol. 29, issue 2, p. 65.
95. Ibid.
96. Wright et al. (2005), *The Physics of Space Security*, p. 165.
97. Ibid.
98. T. N. Barens (2007), 'Miscalculations of Galileo: Europe's Answer to GPS Is Floundering', Center for Defense Information, *Space Security*, 26 September, http://www.cdi.org/program/document.cfm?documentid=4105&programID=68&from_page=../friendlyversion/printversion.cfm, date accessed 3 August 2011.
99. 'What Is EGNOS?'(2011), European Space Agency (ESA) website, http://www.esa.int/esaNA/GGG63950NDC_egnos_0.html, date accessed 7 March 2011.
100. Cf. European Commission (2011), 'A New Space Policy for Europe: Independence, Competitiveness, and Citizens' Quality of Life', *Press Release*, 4 April, http://europa.eu/rapid/pressReleasesAction.do?reference=IP/11/398&type=HTML, date accessed 11 April 2011.
101. European Commission (2010), 'Action Plan on Global Navigation Satellite System (GNSS) Applications', Communication from the Commission to the Council, European Parliament, European Economic and Social Committee and the Committee of the regions, http://eur-lex.europa.eu/LexUriServ/LexUriServ.do?uri=COM:2010:0308:FIN:EN:PDF, date accessed 7 March 2011, p. 3.
102. J. Lembke (2001), 'The Politics of Galileo', *European Policy Paper*, issue 7, European Union Centre, Centre for West European Studies, April, http://www.ucis.pitt.edu/euce/pub/policypapers/2001-Politics_of_Galileo.pdf, date accessed 7 March 2011, p. 19.
103. 'GNSS Applications, Are You Ready? Europe Is!' (2010), European Commission website, http://ec.europa.eu/enterprise/policies/satnav/galileo/files/brochures-leaflets/applications-leaflet_en.pdf, date accessed 7 March 2011.
104. European Commission (2010), 'Action Plan on Global Navigation Satellite System (GNSS) Applications', p. 3.
105. Lembke (2001), 'The Politics of Galileo', p. 30.
106. Ibid., p. 32.
107. M. Sonter (2006), 'Asteroid Mining: Key to the Space Economy', *Space.com*, 9 February, http://www.space.com/adastra/060209_adastra_mining.html, date accessed 8 January 2011.
108. J. Flemming (2011), 'Will Space-Based Solar Take Off?' *Science/Environment*, 5 April, http://www.iewy.com/22074-will-space-based-solar-take-off.html/print/, date accessed 12 April 2011.
109. National Security Space Office (2007), 'Space-Based Solar Power as an Opportunity for Strategic Security', *Report to the Director, National Security*

Space Office, 10 October, http://www.nss.org/settlement/ssp/library/final-sbsp-interim-assessment-release-01.pdf, date accessed 12 April 2011, p. 12.

110. Z. Smeaton (2005), 'Is the Shuttle Green?', *BBC News Magazine*, 8 August, http://news.bbc.co.uk/2/hi/uk_news/magazine/4130980.stm, date accessed 3 August 2011.
111. Book description from editor page about Alexander Bolonkin (2005), *Non-Rocket Space Launch and Flight* (Oxford: Elsevier Science Ltd.), http://www.nextag.com/Non-rocket-Space-Launch-1230268695/specs-html, date accessed 12 April 2011.
112. P. V. Panetta (1998), 'NASA-GSFC Nano-Satellite Technology Development', Paper presented at the 12th AIAA/USU Conference on Small Satellites, 23–6 August 1998, Utah State University, Logan, Utah, http://www.smallsat.org/proceedings/12/ssc98/6/sscvi5.pdf, date accessed 3 August 2011, pp. 1–2.
113. M. Othman (2003), 'Small Satellites for the Benefit of Developing Countries', *Acta Astronautica*, vol. 52, issue 9–12, pp. 687–90.
114. Ibid., p. 690.
115. 'South Africa's First Satellite, SUNSAT', *Cellular Online*, http://www.cellular.co.za/sunsat.htm, date accessed 3 August 2011.
116. Othman (2003), 'Small Satellites for the Benefit of Developing Countries', p. 690.
117. M. Othman (2002), 'Keynote Address: The Small Satellite Business: Contributions to, and from, Developing Countries', in M. J. Rycroft and N. Crosby (eds), *Smaller Satellites: Bigger Business?: Concepts, Applications and Markets for Micro/Nano Satellites in a New Information World* (New York: Springer), p. 2.
118. L. A. Fisk (2008), 'The Impact of Space on Society: Past, Present and Future', *Space Policy*, vol. 24, issue 4, November, p. 180.
119. N. Peter (2009), 'Space Power and Europe in the 21st Century', *European Space Policy Institute Perspectives 21*, April, http://www.espi.or.at/images/stories/dokumente/Perspectives/ESPI_Perspectives_21.pdf, date accessed 27 July 2011, p. 3.

4 Space Security and *Meta*-Geopolitics

1. UNIDIR (2010), 'Space Security 2010: From Foundations to Negotiations', Conference Report 29–30 March (New York and Geneva: United Nations), http://www.unidir.org/pdf/ouvrages/pdf-1-92-9045-010-C-en.pdf, date accessed 26 July 2011, p. 1.
2. M. Krepon and M. Heller (2004), 'A Model Code of Conduct for Space Assurance', *Disarmament Diplomacy*, issue 77, May/June, http://www.acronym.org.uk/dd/dd77/77mkmh.htm, date accessed 3 August 2011, Article I, Definitions.
3. Cf. 'Orbital Debris Frequently Asked Questions' (2009), NASA Orbital Debris Program Office website, http://orbitaldebris.jsc.nasa.gov/faqs.html, date accessed 28 March 2011.
4. T. Sénéchal (2007), 'Space Debris: A Convention Proposal', *Protocol for a Space Debris Risk and Liability Convention*, http://www.pon.org/downloads/ien16.2.Senechal.pdf, date accessed 28 January 2011, p. 41.

5. D. Wright (2007), 'Space Debris', *Physics Today*, vol. 60, issue 8, October, http://www.ucsusa.org/assets/documents/nwgs/wright-space-debris-physics-today.pdf, date accessed 14 March 2011, p. 35.
6. P. Baines and A. Côté (2009), 'Promising Confidence- and Security-Building Measures for Space Security', in UNIDIR, *Disarmament Forum, A Safer Space Environment*, issue 4 (New York and Geneva: United Nations), http://www.unidir.org/bdd/fiche-periodique.php?ref_periodique=1020-7287-2009-4-en, date accessed 3 August 2011, p. 6.
7. Wright (2007), 'Space Debris', pp. 37–8.
8. 'Space Debris: Map', http://maps.thefullwiki.org/Space_debris, date accessed 22 July 2011.
9. NASA (1995), 'Guidelines and Assessment Procedures for Limiting Orbital Debris', *NASA Safety Standard 1740.14*, August, http://orbitaldebris.jsc.nasa.gov/library/NSS1740_14/nss1740_14-1995.html, date accessed 14 August 2011.
10. 'Space Security 2010', *Spacesecurity.org*, August, http://www.spacesecurity.org/space.security.2010.reduced.pdf, date accessed 20 April 2011, p. 35.
11. NASA (2008), *Handbook for Limiting Orbital Debris*, Handbook 8719.14, July, http://www.hq.nasa.gov/office/codeq/doctree/NHBK871914.pdf, date accessed 14 August 2011.
12. R. Walker et.al. (2002), 'Update of the ESA Space Debris Mitigation Handbook', *Executive Summary*, ESA Contract 14471/00/D/HK, July, http://www.esa.int/gsp/completed/execsum00_N06.pdf, date accessed 14 August 2011.
13. 'Space Security 2010', p. 35.
14. Council of the European Union (2010), 'Council Conclusions of 27 September 2010 on the Revised Draft Code of Conduct for Outer Space Activities', 14455/10, 11 October, http://www.consilium.europa.eu/uedocs/cmsUpload/st14455.en10.pdf, date accessed 28 April 2011.
15. 'Space Security 2010', p. 36.
16. Ibid.
17. IADC (2007), 'Space Debris Mitigation Guidelines', IADC-02-01, Revision 1, September, http://www.iadc-online.org/index.cgi?item=docs_pub, date accessed 3 August 2011.
18. United Nations General Assembly (2008), 'International Cooperation in the Peaceful Uses of Outer Space', A/RES/62/217, 1 February, http://www.oosa.unvienna.org/pdf/gares/ARES_62_217E.pdf, date accessed 9 August 2011.
19. 'ESA Space Debris FAQ' (2009), European Space Agency (ESA) website, http://www.esa.int/SPECIALS/Space_Debris/SEM2D7WX3RF_0.html, date accessed 14 June 2011.
20. B. Iannotta and T. Malik (2009), 'US Satellite Destroyed in Space Collision', *Space.com*, 11 February, http://www.space.com/news/090211-satellite-collision.html, date accessed 14 June 2011.
21. Presentation by A. Grebenshchikov at the UNIDIR Space Conference 2009, http://www.unidir.ch/pdf/activites/pdf3act455.pdf, date accessed 14 June 2011, p. 21.
22. 'Space Surveillance - Overview', http://www.globalsecurity.org/space/systems/track-overview.htm, date accessed 14 June 2011.

23. 'Space Situational Awareness Fact Sheet' (2010), http://swfound.org/media/1800/ssa%20fact%20sheet.pdf, date accessed 3 August 2011.
24. 'SDA Overview', Space Data Association website, http://www.space-data.org/sda/about/sda-overview/, date accessed 18 August 2011.
25. Cf. Y. N. Krugly et al. (2010), 'Observations of Asteroids in International Scientific Optical Network', Workshop Gaia Fun-SSO: Follow-up Network for the Solar System Objects in Paris, http://hal.upmc.fr/docs/00/60/28/05/PDF/Krugly-et-ali.pdf, date accessed 14 August 2011.
26. A. Klapovsky, V. Putkov, A. Grebenshchikov and S. Ionov (2009), 'International Exchange of Information on Predictable Dangerous Situations in Outer Space as a New Possible Transparency and Confidence-Building Measure in Space Activities', in UNIDIR, 'Space Security 2009: Moving Toward a Safer Space Environment', Conference 15–16 June, http://www.unidir.ch/pdf/conferences/pdf-conf102.pdf, date accessed 3 August 2011, pp. 1–4.
27. 'Space Security 2010', p. 40.
28. Ibid.
29. Ibid.
30. Ibid.
31. T. Hitchens (2010), 'Multilateralism in Space: Opportunities and Challenges for Achieving Space Security', *Space and Defense*, vol. 4, issue 2, Summer, http://www.unidir.ch/unidir-views/pdf/pdf-uv-29-32.pdf, date accessed 15 August 2011, p. 10.
32. Ibid.
33. ITU Constitution and Convention, http://www.itu.int/net/about/basic-texts/index.aspx, date accessed 15 August 2011.
34. Radio Regulations and International Telecommunication Regulations, http://www.itu.int/net/about/legal.aspx, date accessed 15 August 2011.
35. 'Radiocommunication Sector', International Telecommunication Union (ITU) website, http://www.itu.int/net/about/itu-r.aspx, date accessed 9 June 2011.
36. Ibid.
37. Hitchens (2010), 'Multilateralism in Space: Opportunities and Challenges for Achieving Space Security', *Space and Defense*, p. 12.
38. Ibid.
39. Ibid., p. 11.
40. Cf. F. G. Klotz (1998), *Space, Commerce and National Security* (New York: Council on Foreign Relations), section on The International Politics of Space.
41. ITU (1992), 'Constitution of the International Telecommunication Union', Article 44, http://www.itu.int/net/about/basic-texts/constitution/chaptervii.aspx, date accessed 3 August 2011.
42. Hitchens (2010), 'Multilateralism in Space: Opportunities and Challenges for Achieving Space Security', *Space and Defense*, p. 12.
43. J. Oberg (1998), *Space Power Theory (*Washington, D.C.: Government Printing Office), p. 97.
44. Cf. Klotz (1998), *Space, Commerce and National Security*, section on The International Politics of Space.
45. Hitchens (2010), 'Multilateralism in Space: Opportunities and Challenges for Achieving Space Security', *Space and Defense*, p. 13.

46. 'Space Security 2008', Spacesecurity.org, August, http://www.spacesecurity.org/SSI2008.pdf, date accessed 3 August 2011, p. 41.
47. National Research Council (2008), *Severe Space Weather Events: Understanding Societal and Economic Impacts*, Committee on the Societal and Economic Impacts of Severe Space Weather Events, Workshop Report (Washington, D.C.: National Academies Press), http://www.nap.edu/catalog.php?record_id=12507, date accessed 15 March 2011, pp. 11–12.
48. 'Atmospheric Drag and Lost Objects in Orbit', Windows to the Universe website, http://www.windows2universe.org/spaceweather/lost_sats.html, date accessed 21 January 2010.
49. National Research Council (2008), *Severe Space Weather Events,* p. 11.
50. Ibid., p. 6.
51. A. Lele (2008), 'Militarization of Space', *Indian Defence Review,* vol. 23, issue 2, April–July, http://www.indiandefencereview.com/2008/09/militarization-of-space.html, date accessed 28 March 2011.
52. C. Hoffman (2007), 'Battlefield Space', *Popular Mechanics,* vol. 184, issue 7, July, p. 78.
53. Ibid.
54. L. Grego (2003), 'Short History of US and Soviet ASAT Programs', Union of Concerned Scientists, 1 April, http://www.ucsusa.org/assets/documents/nwgs/asat_history.pdf, date accessed 15 August 2011, p. 1.
55. Ibid., p. 2.
56. F. Tronchetti (2011), 'Preventing the Weaponization of Outer Space: Is a Chinese-Russian-European Common Approach Possible?', *Space Policy,* vol. 27, issue 2, May, p. 83.
57. M. Krepon and S. Black (2009), *Space Security or Anti-Satellite Weapons?* (Washington, D.C.: The Henry L. Stimson Centre), p. 6.
58. 'Anti-satellite weapons (ASATS)', Global Network against Weapons and Nuclear Power in Space website, http://www.space4peace.org/asat/asat.htm, date accessed 15 August 2011.
59. D. Wright (2007), 'Orbital Debris Produced by Kinetic-Energy Anti-Satellite Weapons', in UNIDIR, 'Celebrating the Space Age: 50 Years of Space Technology, 40 Years of the Outer Space Treaty', Conference Report 2–3 April (New York and Geneva: United Nations), http://www.unidir.org/bdd/fiche-article.php?ref_article=2674, date accessed 8 August 2011, p. 158.
60. 'Space Security 2010', p. 157.
61. Ibid.
62. Ibid., p. 158.
63. Ibid.
64. T. A. Summers (2000), 'How Is US Space Power Jeopardized by an Adversary's Exploitation, Technological Developments, Employment and Engagement of Laser Antisatellite Weapons?', Research Report Submitted to the Faculty of the Air Command and Staff College (Alabama: Air University Press Maxwell AFB), http://www.dtic.mil/cgi-bin/GetTRDoc?AD=ADA425117&Location=U2&doc=GetTRDoc.pdf, date accessed 15 August 2011.
65. Ibid., pp. 2–3.
66. Ibid., pp. 12–16.
67. Ibid., pp. 18–19.

68. T. Hitchens (2009), 'Saving Space: Threat Proliferation and Mitigation', Research paper commissioned by the International Commission on Nuclear Non-Proliferation and Disarmament, http://www.icnnd.org/Documents/Hitchens_Saving_Space.pdf?noredirect=1, date accessed 16 August 2011, pp. 8–9.
69. R. Preston, D. J. Johnson, S. J. A. Edwards, M. Miller and C. Shipbaugh (2002), *Space Weapons: Earth War* (Santa Monica: Rand Publications), p. 58.
70. Grego (2003), 'Short History of US and Soviet ASAT Programs', p. 7.
71. J. Singer (2003), 'US-Led Forces Destroy GPS Jamming Systems in Iraq', *Space News*, March, http://utstaging.space.com/spacenews/militaryspace/gps_032503.html, date accessed 28 March 2011.
72. 'Iran Government Jamming Exile Satellite TV' (2005), *Iran Focus*, July, http://www.iranfocus.com/en/?option=com_content&task=view&id=2852; P. de Selding (2007), 'Libya Pinpointed as Source of Months-Long Satellite Jamming in 2006', *Space News*, April, http://www.space.com/3666-libya-pinpointed-source-months-long-satellite-jamming-2006.html, date accessed 28 March 2011.
73. Hitchens (2010), 'Multilateralism in Space: Opportunities and Challenges for Achieving Space Security', *Space and Defense*, p. 14.
74. Ibid.
75. P. Piazza (2003), 'Cyberthreats to Satellite Security', *Security Management*, vol. 47, issue 2, p. 34.
76. United States General Accounting Office (GAO) (2002), 'Critical Infrastructure Protection: Commercial Satellite Security Should Be More Fully Addressed', *Report to the Ranking Minority Member Permanent Subcommittee on Investigations, Committee on Governmental Affairs US Senate*, August, http://www.gao.gov/new.items/d02781.pdf, date accessed 28 March 2011, p. 2.
77. J. A. Lewis (2007), 'China's Military Modernization and Its Impact on the United States and Asia Pacific Region', Testimony before the US-China Economic and Security Review Commission, 30 March, http://csis.org/files/media/csis/congress/ts070330lewis.pdf, date accessed 14 June 2011, p. 5.
78. Ibid.
79. United States GAO (2002), 'Critical Infrastructure Protection: Commercial Satellite Security Should be More Fully Addressed', p. 1.
80. Lewis (2007), 'China's Military Modernization and Its Impact on the United States and Asia Pacific Region', p. 6.
81. Ibid.
82. United States GAO (2002), 'Critical Infrastructure Protection: Commercial Satellite Security Should be More Fully Addressed', p. 1.
83. Ibid., p. 2.
84. J. Oberg (2007), 'The Dozen Space Weapons Myths', *Space Review*, 12 March, http://www.thespacereview.com/article/826/1, date accessed 18 August 2011.
85. B. M. DeBlois, R. L. Garwin, S. Kemp and J. C. Marwell (2005), 'Star Crossed', *IEEE Spectrum*, vol. 42, issue 3, March, pp. 40–9.

86. Cf. J. Cirincione (2000), 'Brief History of Ballistic Missile Defense and Current Programs in the United States', adapted from 'The Persistence of the Missile Defense Illusion', Paper presented at a conference in Como, Italy, 2–4 July 1998, http://www.carnegieendowment.org/publications/index.cfm?fa=view&id=133&prog=zgp&proj=znpp#starwars, date accessed 28 March 2011.
87. M. Krepon (2004), 'Weapons in the Heavens: A Radical and Reckless Option', *Arms Control Today*, November, http://www.armscontrol.org/act/2004_11/Krepon, date accessed 29 March 2011.
88. D. Thisdell (2011), 'Space Junk Raises Alarm', *Flight International*, 13 April, http://www.flightglobal.com/articles/2011/04/13/355537/space-junk-raises-alarm.html, date accessed 18 April 2011.
89. European Commission (2002), *The European Dependence on US-GPS and the GALILEO Initiative* (Brussels: European Commission), http://galileo.khem.gov.hu/documents/angol/technikai_dokumentumok/the_european_dependence_on_gps_and_the_galileo_initiative.pdf, date accessed 28 March 2011.
90. Cf. E. Morris (2008), 'A Day Without Space', Remarks during the Public Discussion Forum, US Chamber of Commerce, http://www.space.commerce.gov/library/speeches/2008-10-daywospace.shtml, date accessed 15 March 2011.
91. 'Telemedicine: Health Alert via Satellite' (2007), European Space Agency (ESA) website, 30 November, http://www.esa.int/esaTE/SEM7DK73R8F_index_0.html, date accessed 28 March 2011.
92. Cf. Morris (2008), 'A Day without Space'.
93. M. Hapgood (2011), 'Space Weather: Nature's Electromagnetic Hazard', *102nd Kelin Lecture*, March, http://tv.theiet.org/technology/communications/10604.cfm, date accessed 14 March 2011.
94. National Research Council (2008), *Severe Space Weather Events*, p. 6.
95. Ibid., p. 2.
96. Ibid.
97. Ibid., Preface, p. vii.
98. Ibid.
99. Ibid.
100. Ibid.
101. Ibid., p. 1.
102. Ibid., p. 3.
103. Ibid., p. 33.
104. D. Wright, L. Grego and L. Gronlund (2005), *The Physics of Space Security: A Reference Manual* (Cambridge, MA: American Academy of Arts and Sciences), http://www.amacad.org/publications/Physics_of_Space_Security.pdf, date accessed 4 August 2011, p. 169.
105. 'National Missile Defense' (2000), Space Policy Project Special Weapons Monitor Section, Federation of American Scientists (FAS) website, June, http://www.fas.org/spp/starwars/program/nmd/, date accessed 28 March 2011.
106. C. V. Peña and E. L. Hudgins (2002), 'Should the United States "Weaponize" Space? Military and Commercial Implications', *Policy Analysis*, issue 427, March, p. 8.

107. Ibid.
108. Ibid.
109. R. Weitz (2007), 'Special Report: Chinese Anti-Satellite Weapon Test, the Shot Heard Round the World', *WMD Insight*, March, http://www.wmdinsights.com/I13/I13_EA1_SP_II_RussiaChinese.htm, date accessed 28 March 2011, Part I: US Allies Criticize China's Anti-Satellite Weapon Test; Media Notes Concerns About US Space Policies.
110. Thisdell (2011), 'Space Junk Raises Alarm', *Flight International*.
111. C. Zissis (2007), 'China's Anti-Satellite Test', *Council on Foreign Relations*, 22 February, http://www.cfr.org/publication/12684/chinas_antisatellite_test.html#1, date accessed 28 March 2011.
112. B. W. MacDonald (2008), 'China, Space Weapons, and US Security', *Council Special Report,* issue 38, September, http://www.cfr.org/content/publications/attachments/China_Space_CSR38.pdf, date accessed 29 March 2011, p. 3.
113. E. Morris, S. Anderson, R. Hatch, P. Hays, J. Armor and J. Sheldon (2008), 'A Day Without Space: Economic and National Security Ramifications', *Washington Roundtable on Science and Public Policy* (Washington, D.C.: The George Marshall Institute), http://www.marshall.org/pdf/materials/695.pdf, date accessed 4 August 2011, p. 3.
114. Krepon (2004), 'Weapons in the Heavens: A Radical and Reckless Option'.
115. Ibid.
116. N. R. F. Al-Rodhan (2007), *The Five Dimensions of Global Security: Proposal for a Multi-sum Security Principle* (Zurich: LIT), p. 31.
117. Ibid.
118. United Nations Development Programme (1994), *New Dimensions of Human Security* (New York: UNDP).
119. UNISPACE 3 (1999), 'Conference Urges Creation of Voluntary Fund to Implement Its Recommendations, Further Work on Legal Aspects of Space Debris, Support for Regional Education Centres', Third United Nations Conference on the Exploration and Peaceful Uses of Outer Space, Vienna Declaration on Space and Human Development, 30 July, http://www.un.org/events/unispace3/pressrel/e30pm.htm, date accessed 28 January 2011.
120. Al-Rodhan (2007), *The Five Dimensions of Global Security,* p. 60.
121. J. A. Lewis (2007), 'Space and National Security', *NEC Space and Defense Forum*, Tokyo, November, http://csis.org/files/media/csis/pubs/071115_nec.pdf, date accessed 29 March 2011, p. 1.
122. Al-Rodhan (2007), *The Five Dimensions of Global Security*, pp. 72–3.
123. N.-L. Remuss (2009), 'The Need to Counter Space Terrorism: A European Perspective', *European Space Policy Institute Perspectives*, issue 17, January, http://www.espi.or.at/images/stories/dokumente/Perspectives/espi%20perspectives%2017.pdf, date accessed 4 February 2010, p. 1.
124. Ibid., p. 3.
125. Xinhua (2007), 'Chinese Satellite TV Hijacked by Falun Gong Cult', *Facts.org,* 12 July, http://www.facts.org.cn/Feature/fgths/200712/t72127.htm, date accessed 27 January 2011.
126. Al-Rodhan (2007) *The Five Dimensions of Global Security,* p. 78.
127. N. Jasentuliyana and K. Karnik (1997), 'Space Futures and Human Security', *Space Policy*, vol. 13, issue 3, August, p. 260.

128. Ibid., p. 260.
129. Remuss (2009), 'The Need to Counter Space Terrorism: A European Perspective', p. 3; J. Daly (2007) 'LTTE: Technologically Innovative Rebels', *Spero News*, 5 June, http://www.speroforum.com/a/9803/LTTE-Technologically-innovative-rebels, date accessed 27 January 2011.
130. L. Grego and D. Wright (2010), *Securing the Skies: Ten Steps the United States Should Take to Improve the Security and Sustainability of Space* (Washington, D.C.: Union of Concerned Scientists), http://www.ucsusa.org/assets/documents/nwgs/securing-the-skies-full-report-1.pdf, date accessed 31 March 2011, p. 6.
131. S. Ordzhonikidze cited in UNIDIR (2010), 'Space Security 2010: From Foundations to Negotiations', Conference Report 29–30 March (New York and Geneva: United Nations), pp. 2–3.
132. J. C. Moltz (2008), *The Politics of Space Security: Strategic Restraint and the Pursuit of National Interests* (Stanford: Stanford University Press), pp. 23–41.
133. P. L. Hays (2002), 'Current and Future Military Uses of Space', Outer Space and Global Security Conference, Geneva, 26–27 November, http://www.unidir.org/pdf/articles/pdf-art1989.pdf, date accessed 4 August 2011, pp. 35–8.
134. J.-M. Stoullig (2011), 'Rumsfeld Commission Warns Against "Space Pearl Harbor"', *Space Daily*, 11 January, http://www.spacedaily.com/news/bmdo-01b.html, date accessed 4 August 2011.
135. Cf. R. Johnson (2002), 'Space Security: Options and Approaches', Paper Presented at the Outer Space and Global Security Conference, Geneva, 26–7 November 2002.
136. J. Kueter (2008), 'Space Treaty Would Hurt US', *USA Today*, 21 February, http://www.usatoday.com/printedition/news/20080221/oppose21.art.htm, date accessed 29 March 2011.
137. DeBlois (1998), 'Space Sanctuary: A Viable National Strategy', *Airpower Journal*, vol. 12, no. 4, pp. 51–2.
138. T. Hitchens (2008), 'Space Wars', *Scientific American*, vol. 298, issue 3, 1 March, p. 85.
139. B. M. DeBlois (2003), *Outer Space and Global Security* (Geneva: United Nations Publications), p. 19.
140. Since the signing of the Outer Space Treaty in 1967, this framework has grown to include the Astronaut Rescue Agreement (1968), the Liability Convention (1972), the Registration Convention (1979) and the Moon Agreement (1979) as well as a range of other international and bilateral agreements and relevant rules of customary international law.
141. Baines and Côté (2009), 'Promising Confidence- and Security-Building Measures for Space Security', in UNIDIR (2009), *Disarmament Forum: A Safer Space Environment*, p. 5.
142. A. B. Englehart (2008), 'Common Ground in the Sky: Extending the 1967 Outer Space Treaty to Reconcile U.S. and Chinese Security Interests', Pacific Rim Law & Policy Journal, vol. 17, issue 1, pp. 134, 142.
143. Baines and Côté (2009), 'Promising Confidence- and Security-Building Measures for Space Security', p. 8.

144. 'Executive Summary', in *Space Security 2007*, Spacesecurity.org, August, http://www.spacesecurity.org/SSI2007ExecutiveSummary.pdf, date accessed 29 March 2011, p. 9.
145. International Security Bureau of the Department of Foreign Affairs and International Trade (2004), *Space Security 2003*, Research Report prepared for the International Security Bureau of the Department of Foreign Affairs and International Trade, http://www.international.gc.ca/arms-armes/assets/pdfs/space-espace.pdf, date accessed 29 March 2011, p. 10.
146. *Space Security 2008*, Spacesecurity.org, August, http://www.spacesecurity.org/SSI2008.pdf, date accessed 29 March 2001, p. 3.
147. Ibid., p. 4.
148. 'China Establishes First Space Trash Observation Center' (2005), *China View*, 31 November, http://news.xinhuanet.com/english/2005-03/11/content_2682880.htm, date accessed 15 August 2011.
149. Al-Rodhan (2007), *The Five Dimensions of Global Security*, pp. 129–30.
150. N. R. F. Al-Rodhan (2009), *Neo-Statecraft and Meta-Geopolitics*. Reconciliation of Power, Interests and Justice in the 21st Century (Berlin: LIT), p. 54.

5 Space Power and Applied *Meta*-Geopolitics

1. N. Peter (2008), 'Space Power and Europe: In the Need for a Conceptual Framework', *Proceedings of the IAC, International Astronautical Congress*, http://www.espi.or.at/images/stories/dokumente/presentations/2008/iac-08-e3.2.9-nicolas_peter.pdf, date accessed 27 July 2011 p. 2.
2. Ibid., p. 3.
3. Ibid.
4. N. Peter (2009), 'Space Power and Europe in the 21st Century', *European Space Policy Institute Perspectives*, issue 21, April, http://www.espi.or.at/images/stories/dokumente/Perspectives/ESPI_Perspectives_21.pdf, date accessed 27 July 2011, p. 1.
5. W. J. Lynn, III (2011), 'A Military Strategy for the New Space Environment', *Washington Quarterly*, vol. 32, issue 3, Summer, http://www.twq.com/11summer/index.cfm?id=440, date accessed 6 July 2011, p. 9.
6. US DOD and ODNI (2011), *National Security Space Strategy: Unclassified Summary*, January, http://www.defense.gov/home/features/2011/0111_nsss/docs/NationalSecuritySpaceStrategyUnclassifiedSummary_Jan2011.pdf, date assessed 24 March 2011, p.1.
7. Ibid.
8. Peter (2009), 'Space Power and Europe in the 21st Century', p. 2.
9. N. R. F. Al-Rodhan (2009), *Neo-Statecraft and Meta-Geopolitics: Reconciliation of Power, Interests, and Justice in the 21st Century* (Berlin: LIT), p. 202.
10. M. S. Smith (2011), 'President Obama's National Space Policy: A Change in Tone and a Focus on Space Sustainability', *Space Policy*, vol. 27, issue 1, p. 20.
11. J. Foust (2010), 'A Change in Tone in National Space Policy', *Space Review*, 6 July, http://www.thespacereview.com/article/1660/1, date accessed 31 March 2011.
12. C. Homans (2010), 'The Wealth of Constellations', *Washington Monthly*, May–June, http://www.washingtonmonthly.com/features/2010/1005.homans.html, date accessed 7 June 2011.

13. Ibid.
14. White House (2010), *National Space Policy of the United States of America*, 28 June, http://www.whitehouse.gov/sites/default/files/national_space_policy_6-28-10.pdf, date accessed 25 March 2011, p. 4.
15. Ibid.
16. L. Grego and D. Wright (2010), *Securing the Skies: Ten Steps the United States Should Take to Improve the Security and Sustainability of Space* (Washington, D.C.: Union of Concerned Scientists), http://www.ucsusa.org/assets/documents/nwgs/securing-the-skies-full-report-1.pdf, date accessed 31 March 2011, p. 1.
17. J. D. Rendleman and J. W.Faulconer (2010), 'Improving International Space Cooperation: Considerations for the USA', *Space Policy*, vol. 26, issue 3, pp. 144–5.
18. Grego and Wright (2010), *Securing the Skies*, p. 1.
19. Ibid.
20. Space Foundation (2011), 'The Authoritative Guide to Global Space Activity', *The Space Report 2011: Executive Summary*, http://www.thespacereport.org/files/The_Space_Report_2011_exec_summary.pdf, date accessed 25 May 2011, p. 8.
21. Grego and Wright (2010), *Securing the Skies*, p. 10.
22. Ibid.
23. White House (2010), *National Space Policy of the United States of America*, p. 3.
24. Ibid., p. 5.
25. Ibid., p. 7.
26. Foust (2010), 'A Change in Tone in National Space Policy'.
27. Grego and Wright (2010), *Securing the Skies*, p. 10.
28. B. Sutherland (2009), 'Why America Is Lost in Space', *Newsweek*, 30 January, http://www.newsweek.com/2009/01/30/why-america-is-lost-in-space.html, date accessed 5 April 2011.
29. Ibid.
30. Grego and Wright (2010), *Securing the Skies*, p. 10.
31. G. L. Schulte (2011), 'National Space Security Strategy Outlines Rules of the Road', *Globalsecurity.org*, February, http://www.globalsecurity.org/space/library/news/2011/space-110211-afps01.htm, date accessed 31 March 2011.
32. Ibid.
33. Homans (2010), 'The Wealth of Constellations'.
34. Ibid.
35. Foust (2010), 'A Change in Tone in National Space Policy'.
36. R. Weitz (2011), 'New US National Security Space Strategy Proposes New Partnerships', Second Line of Defense (SLD) website, 22 March, http://www.sldinfo.com/?p=16425, date accessed 1 April 2011.
37. Space Foundation (2011), 'The Authoritative Guide to Global Space Activity', The Space Report 2011: Executive Summary, p. 8.
38. White House (2010), *National Space Policy of the United States of America*, p. 6.
39. 'Space Security 2010', *Spacesecurity.org*, August, http://www.spacesecurity.org/space.security.2010.reduced.pdf, date accessed 20 April 2011, p. 74.
40. R. Skaar (2007), 'Commercialisation of Space and its Evolution: Will New Ways to Share Risks and Benefits Open Up a Much Larger Space Market?',

European Space Policy Institute (ESPI) Report 4, May, http://www.espi.or.at/images/stories/dokumente/studies/commercialisation.pdf, date accessed 4 April 2011, p. 27.

41. Weitz (2011), 'New US National Security Space Strategy Proposes New Partnerships'.
42. E. Hagt (2006), 'Mutually Assured Vulnerabilities in Space,' *China Security*, issue 2, 21 March, http://www.wsichina.org/space/attach/CS2_8.pdf, date accessed 29 April 2011, p. 90.
43. C. V. Peña and E. L. Hudgins (2002), 'Should the United States "Weaponize" Space? Military and Commercial Implications', *Policy Analysis*, issue 427, March, p. 16.
44. US DOD and ODNI (2011), *National Security Space Strategy*, pp. 1–3.
45. Peña and Hudgins (2002), 'Should the United States "Weaponize" Space?', p. 8.
46. Smith (2011), 'President Obama's National Space Policy', p. 20.
47. Ibid.
48. Lynn (2011), 'A Military Strategy for the New Space Environment', p. 10.
49. Ibid., p. 11.
50. Ibid., p. 12.
51. M. Krepon, T. Hitchens and M. Katz-Hyman (2011), 'Preserving Freedom of Action in Space: Realizing the Potential and Limits of US Spacepower', in C. D. Lutes, P. L. Hays (eds), with V. A. Manzo, L. M. Yambric, M. E. Bunn, *Toward a Theory of Spacepower: Selected Essays* (Washington, D.C.: National Defense University Press), http://www.ndu.edu/press/spacepower.html, date accessed 6 June 2011, Chapter 20, p. 395.
52. White House (2010), *National Space Policy of the United States of America*, pp. 6–7.
53. Grego and Wright (2010), *Securing the Skies*, pp. 8–9.
54. Ibid., p. 9.
55. G. L. Schulte (2011), 'The Final Frontier', *Foreign Policy*, 9 February, http://www.foreignpolicy.com/articles/2011/02/09/the_final_frontier, date accessed 4 August 2011.
56. J. D. Rendleman and J. W. Faulconer (2010), 'Improving International Space Cooperation: Considerations for the USA', *Space Policy*, vol. 26, issue 3, p. 144.
57. Smith (2011), 'President Obama's National Space Policy', p. 22.
58. 'What's the Future of US-China Cooperation in Space?' (2011), *Space Politics*, 5 May, http://www.spacepolitics.com/2011/05/05/whats-the-future-of-us-china-cooperation-in-space/, date accessed 19 August 2011.
59. The House of Representatives (2011), 'H. R. 1473. Bill', *FY 2011 Continuing Resolution*, 11 April, http://www.gpo.gov/fdsys/pkg/BILLS-112hr1473ih/pdf/BILLS-112hr1473ih.pdf, date accessed 19 August 2011, p. 215.
60. Ibid.
61. Rendleman and Faulconer (2010), 'Improving International Space Cooperation', p. 149.
62. Peter (2009), 'Space Power and Europe in the 21st Century', p. 4.
63. Grego and Wright (2010), *Securing the Skies*, p. 1.
64. J. Kueter (2010), 'Evaluating the Obama National Space Policy: Continuity and New Priorities', *George C. Marshall Institute Policy Outlook*, July, http://www.marshall.org/pdf/materials/900.pdf, date accessed 1 April 2011, p. 11.

65. Ibid.
66. Antonio Tajani expressed his opinion in European Commission (2011), 'A New Space Policy for Europe: Independence, Competitiveness and Citizen's Quality of Life', Press Release, Brussels, 4 April, http://europa.eu/rapid/pressReleasesAction.do?reference=IP/11/398&format=HTML&aged=0&language=EN&guiLanguage=en, date accessed 20 April 2011.
67. S. P. Worden (2011), 'Future Strategy and Professional Development: A Roadmap', in Lutes and Hays (eds), et al. (2011), *Toward a Theory of Spacepower*, p. 578.
68. European Commission (2009), 'Europe, a Global Actor in Space', *Brochure*, http://ec.europa.eu/enterprise/policies/space/files/policy/entr_brochspace_en.pdf, date accessed 20 April 2011.
69. S. Pagkratis (2011), *Space Policies, Issues and Trends in 2010/2011*, ESPI Report 35, June (Vienna: ESPI), p. 40.
70. European Commission (2011), 'Towards a Space Strategy for the European Union that Benefits Its Citizens', *Communication from the Commission to the Council, the European Parliament, The European Economic and Social Committee and the Committee of the regions*, COM/2011/152, Brussels, 4 April, http://ec.europa.eu/enterprise/policies/space/files/policy/comm_pdf_com_2011_0152_f_communication_en.pdf, date accessed 20 April 2011, p. 8.
71. Rendleman and Faulconer (2010), 'Improving International Space Cooperation', pp. 144–5.
72. L. Nardon and C. Venet (2011), 'GMES, the Second Flagship', *The Europe & Space Series*, issue 3, March, http://www.ifri.org/?page=detail-contribution&id=6465, date accessed 10 June 2011, p. 1.
73. European Commission (2009), 'Europe, a Global Actor in Space'.
74. Ibid.
75. Peter (2009), 'Space Power and Europe in the 21st Century', pp. 3–4.
76. 'Space Security 2010', pp. 71–2.
77. 'Taking Europe into the 21st Century', European Union website, http://europa.eu/lisbon_treaty/index_en.htm, date accessed 21 April 2011.
78. European Union (2007), 'Treaty of Lisbon Amending the Treaty on European Union and the Treaty Establishing the European Community', 13 December, 2007/C 306/01, http://eurlex.europa.eu/JOHtml.do?uri=OJ:C:2007:306:SOM:EN:HTML, date accessed 21 April 2011, Art 189.
79. Peter (2009), 'Space Power and Europe in the 21st Century', p. 5.
80. European Commission (2011), 'A New Space Policy for Europe: Independence, Competitiveness and Citizen's Quality of Life', *Press Release*, Brussels, 4 April, http://europa.eu/rapid/pressReleasesAction.do?reference=IP/11/398&format=HTML&aged=0&language=EN&guiLanguage=en, date accessed 20 April 2011.
81. European Commission (2011), 'Towards a Space Strategy for the European Union that Benefits Its Citizens', p. 3.
82. Ibid.
83. Ibid.
84. 'What is EGNOS?', ESA website, http://www.esa.int/esaNA/GGG63950NDC_egnos_0.html, date accessed 5 August 2011.
85. 'GMES: Observing Our Planet for a Safer World', ESA website, http://ec.europa.eu/enterprise/policies/space/gmes/, date accessed 5 August 2011.

86. For detailed information about *Europe 2020: A Strategy for a Smart, Sustainable and Inclusive Growth* (COM (2010) 2020), see 'Europe 2020', European Commission website, http://ec.europa.eu/europe2020/index_en.htm, date accessed 5 August 2011.
87. For detailed information about *An Integrated Industrial Policy for the Globalization Era-Putting Competitiveness and Sustainability at the Centre Stage* (COM (2010) 614), see 'Europe 2020 Flagship: An Industrial Policy for the Globalisation Era', European Commission website, http://ec.europa.eu/enterprise/policies/industrial-competitiveness/industrial-policy/index_en.htm, date accessed 5 August 2011.
88. European Commission (2011), 'Towards a Space Strategy for the European Union that Benefits Its Citizens', p. 8.
89. 'European Space Agency' (2007), *ESA Bulletin*, http://www.esa.int/esapub/bulletin/bulletin131/bul131_colophon.pdf, date accessed 26 April 2011.
90. M. Sheehan (2007), 'European Integration and Space', *The International Politics of Space* (London: Routledge), p. 85.
91. Ibid.
92. 'Industrial Policy and Geographical Distribution' (2007), *How to do Business with ESA*, Topic 7, ESA website, http://www.esa.int/esaMI/industry_how_to_do_business/SEMDWN6H07F_0.html, date accessed 26 April 2011.
93. European Commission (2011), 'Towards a Space Strategy for the European Union that Benefits Its Citizens', p. 8.
94. 'Space Foundations', European Commission website, http://ec.europa.eu/enterprise/policies/space/esp/foundations/index_en.htm, date accessed 8 August 2011.
95. Sheehan (2007), 'European Integration and Space', pp. 83–5.
96. The author refers to two precursor organizations: the European Launcher Development Organization (ELDO) and European Space Research Organization (ESRO), both of which merged into the ESA in 1975. For more details, see M. Sheehan (2007), 'European Integration and Space', *The International Politics of Space* (London: Routledge), pp. 72–90.
97. Ibid., p. 84.
98. European Commission (2011), 'Towards a Space Strategy for the European Union that Benefits Its Citizens', p. 12.
99. Ibid., p. 7.
100. Ibid.
101. Ibid., pp. 5–6.
102. M. Sheehan (2009), 'Profaning the Path to the Sacred. The Militarization of the European Space Programme', in N. Bormann, M. Sheehan (eds) *Securing Outer Space* (New York and Milton: Routledge), p. 175.
103. Ibid., p. 177.
104. N. R. F. Al-Rodhan (2007), *The Five Dimensions of Global Security: Proposal for a Multi-sum Security Principle* (Zurich: LIT), p. 35.
105. Sheehan (2009), 'Profaning the Path to the Sacred', p. 179.
106. European Commission (2003), 'Space: A New European Frontier for an Expanding Union. An Action Plan for Implementing the European Space Policy', White Paper, COM /200/673, 11 November (Luxembourg: European Communities), p. 19.

107. Formerly known as the European Security and Defence Policy (ESDP). The Treaty of Lisbon renamed the ESDP the Common Security and Defence Policy (CSDP).
108. 'What We Do', European Defence Agency (EDA) website, http://www.eda.europa.eu/Aboutus/Whatwedo, date accessed 27 April 2011.
109. Cf. EDA (2006), 'An Initial Long-term Vision for European Defence Capability and Capacity Needs', *Report of the European Defence Agency*, issue 2006/11, 3 October, http://www.eda.europa.eu/webutils/downloadfile.aspx?fileid=105, date accessed 27 April 2011.
110. Cf. JAPCC (2009), *NATO Space Operations Assessment* (Kalkar, Germany: Joint Air Power Competence Centre), January, http://www.japcc.de/108.html, date accessed 5 August 2011, pp. 1–53.
111. S. J. Frontinus (2009), 'NATO Space Operations Today', in JAPCC (2009), *NATO Space Operations Assessment* pp. 21–31.
112. Commission of the European Communities (2007), 'European Space Policy', *Communication from the Commission to the Council and the European Parliament,* COM(2007) 212 final, 26 April, Brussels, http://eur-lex.europa.eu/LexUriServ/LexUriServ.do?uri=COM:2007:0212:FIN:en:PDF, date accessed 5 August 2011; Council of the European Union (2007), *Resolution on the European Space Policy*, 10037/07, 25 May, Brussels,http://www.smm.lt/smt/docs/tm_tyrimai/esmtep/st10037%20Resolution%20on%20ESP%20as%20adopted%20in%20EC%20CC%20%20en2007-05-25.pdf, date accessed 5 August 2011.
113. Ibid.
114. M. Dickow (2007), 'Security and Defence in the European Space Policy, *ESPI Flash Report*, issue 2, June, http://www.espi.or.at/images/stories/dokumente/flash_reports/flash-report2-espi-esdp-june2007.pdf, date accessed 5 August 2011, p. 3.
115. Ibid.
116. European Commission (2011), 'Towards a Space Strategy for the European Union that Benefits Its Citizens', p. 6.
117. The Space Council cited in European Commission (2011), 'Towards a Space Strategy for the European Union that Benefits Its Citizens', p. 6.
118. Nardon and Venet (2011), 'GMES: The Second Flagship'.
119. Ibid.
120. Ibid.
121. Ibid.
122. B. James (2001), 'Washington Said to Fear Use of Galileo by Enemy in a War: US Out of Line on Global Positioning, EU Says', *New York Times*, 19 December, http://www.nytimes.com/2001/12/19/news/19iht-gps_ed3_.html?scp=2&sq=19%20December%202001%20jaques%20chirac&st=cse, date accessed 27 April 2011.
123. F. Slijper (2009), 'The EU Should Freeze Its Military Ambitions in Space', *Space Policy*, vol. 25, p. 72.
124. European Parliament (2008), 'European Parliament Resolution on Space and Security' (2008/2030(INI), 10 July, http://www.europarl.europa.eu/sides/getDoc.do?type=TA&reference=P6-TA-2008-0365&language=EN., date accessed 5 August 2011, General Consideration 12.

125. Ibid., General Consideration 44.
126. European Commission (2011), 'Towards a Space Strategy for the European Union that Benefits Its Citizens', p. 2.
127. Ibid., p. 6.
128. Ibid., p. 7.
129. 'What is SSA?'(2010), ESA website, http://www.esa.int/esaMI/SSA/SEMYTICKP6G_0.html, date accessed 27 April 2011.
130. 'SSA: Five Questions with ESA's Nicolas Bobrinsky' (2008), ESA website, 13 November, http://www.esa.int/SPECIALS/SSA/SEMFSG6EJLF_1_iv.html, date accessed 27 April 2011.
131. 'Russian and US Satellites Collide' (2009), *BBC News*, 12 February, http://news.bbc.co.uk/2/hi/7885051.stm, date accessed 8 August 2011.
132. 'A New Space Policy for Europe' (2011), *Space Daily*, 7 April, http://www.spacedaily.com/reports/A_New_Space_Policy_For_Europe_999.html, date accessed 28 April 2011.
133. N. Peter (2010), 'Space Power and Its Implications: The Case of Europe', *Acta Astronautica,* vol. 66, issues 3–4, February/March, p. 354.
134. P. B. de Selding (2011), 'European Commission Urges China Dialogue', *Space News*, 5 April, http://www.spacenews.com/civil/110405-space-policy-document-calls-for-dialogue-with-china.html, date accessed 28 April 2011.
135. Ibid.
136. 'Space Security 2010', p. 130.
137. Sheehan (2007), 'European Integration and Space', pp. 89–90.
138. K.-U. Schrogl (2008), 'The European Space Policy: Still in Need of a Regulatory Framework for European Space Activities', *Aviation and Maritime Journal*, issue 2, Anno VII, October/December, http://www.espi.or.at/images/stories/dokumente/press/the%20aviation%20%26%20maritime%20journal%20n.%204-2008.pdf, date accessed 28 April 2011, p. 2.
139. B. De Montluc (2009), 'The New International Political and Strategic Context for Space Policies', *Space Policy*, vol. 25, issue 1, February, p. 27.
140. Peter (2010), 'Space Power and Its Implications: The Case of Europe', p. 353.
141. EU Council (2008), 'Draft Code of Conduct for Outer Space Activities', 17175/08, PESC 1697, CODUN 61, Brussels, 17 December, http://register.consilium.europa.eu/pdf/en/08/st17/st17175.en08.pdf, date accessed 8 August 2011.
142. W. Rathgeber, N. Remuss and K. Schrogl (2009), 'Space Security and the European Code of Conduct for Outer Space Activities', in UNIDIR, *Disarmament Forum: A Safer Space Environment*, issue 4 (New York and Geneva: United Nations), http://www.unidir.org/pdf/articles/pdf-art2909.pdf, date accessed 20 April 2011, p. 34.
143. Council of the European Union (2010), 'Council Conclusions of 27 September 2010 on the Revised Draft Code of Conduct for Outer Space Activities', 11 October 2010, http://www.consilium.europa.eu/uedocs/cmsUpload/st14455.en10.pdf, date accessed 28 April 2011, p. 8.

144. J. Foust (2011), 'Debating a Code of Conduct for Space', *Space Review*, 7 March, http://www.thespacereview.com/article/1794/1, date accessed 28 April 2011.
145. Rathgeber, Remuss and Schrogl (2009), 'Space Security and the European Code of Conduct for Outer Space Activities', in UNIDIR (2009), *Disarmament Forum: A Safer Space Environment*, p. 40.
146. de Selding (2011), 'European Commission Urges China Dialogue'.
147. Ibid.
148. The GMES is jointly funded by the EU and the ESA, but the funding for *Sentinel* satellites after 2013 is not guaranteed. As regards Galileo, the European Commission announced in January 2010 a 59 per cent budget overrun, mostly due to the increase in the price of the Russian launchers and initial underestimates of the costs of adapting Soyuz rockets to the Galileo mission. S. Pagkratis (2010), 'Space Policies, Issues and Trends in 2009/2010', *ESPI Report*, issue 23, June, http://www.isn.ethz.ch/isn/Digital-Library/Publications/Detail/?id=124783, date accessed 2 May 2011, pp. 29, 32.
149. M. Meyer (2011), 'Political Advance Hangs on Post-2013 Financial Perspectives', *Europolitics/Research*, 4 April, http://www.europolitics.info/political-advance-hangs-on-post-2013-financial-perspectives-art300421-1.html, date accessed 28 April 2011.
150. S. Pagkratis (2010), 'Space Policies, Issues and Trends in 2009/2010', *ESPI Report*, issue 23, June, http://www.isn.ethz.ch/isn/Digital-Library/Publications/Detail/?id=124783, date accessed 2 May 2011, p. 52.
151. G. B. Kaplan, J. M. Telep, T. K. Williams and S. B. Wilson (2011), 'China's 12th Five-Year Plan for National Economic and Social Development Provides a Glimpse of China's Industrial Focus in the Near Term', *Martindale.com*, 7 April, http://www.martindale.com/government-law/article_King-Spalding-LLP_1265284.htm, date accessed 5 May 2011.
152. Information Office of the State Council of the People's Republic of China (2006), 'China's Space Activities in 2006', *White Paper*, 12 October, http://www.fas.org/spp/guide/china/wp2006.pdf, date accessed 5 August 2011, Section II – Progress made in the past five years.
153. J. P. Acuthan (2006), 'China's Outer Space Programme: Diplomacy of Competition or Cooperation?' *China Perspectives*, issue 63, January/February, p. 3; S. Writers (2006), 'Chinese Annual Space Budget Exceeds Two Billion Dollars', *Space Daily*, 12 October, http://www.spacedaily.com/reports/Chinese_Annual_Space_Budget_Exceeds_Two_Billion_Dollars_999.html, date accessed 15 February 2011.
154. J. Johnson-Freese (2007), 'China's Space Ambitions', *Ifri Proliferation Papers*, issue 18, Summer, http://www.ifri.org/?page=detail-contribution&id=4833&id_provenance=97, date accessed 3 May 2011, p. 13.
155. Information Office of the State Council of the People's Republic of China (2006), 'China's Space Activities in 2006', Section I.
156. K. Pollpeter (2008), *Building for the Future: China's Progress in Space Technology During the Tenth 5-Year Plan and the US Response*, Strategic Studies Institute, March, http://www.strategicstudiesinstitute.army.mil/pdffiles/PUB852.pdf, date accessed 3 May 2011, p. 28.

157. Rendleman and Faulconer (2010), 'Improving International Space Cooperation, pp. 144–5.
158. Acuthan (2006), 'China's Outer Space Programme: Diplomacy of Competition or Cooperation?', p. 8.
159. Information Office of the State Council of the People's Republic of China (2006), 'China's Space Activities in 2006', Section IV – Development policies and measures.
160. A. De Neve (2011), 'Reconsidering the Military Axis of the Chinese Space Programme', *ESPI Perspectives*, issue 48, 28 June, http://www.espi.or.at/index.php?option=com_content&view=article&id=704:espi-perspective-on-the-security-aspects-of-the-chinese-space-programme-on-line&catid=39:news-archive&Itemid=37, date accessed 8 July 2011, p. 1.
161. Ibid., p. 4.
162. Ibid.
163. C. Mathieu (2008), 'Assessing Russia's Space Cooperation with China and India: Opportunities and Challenges for Europe', *ESPI Report,* issue 12, June, http://www.isn.ethz.ch/isn/Digital-Library/Publications/Detail/?ots591=cab359a3-9328-19cc-a1d2-8023e646b22c&lng=en&id=124767, date accessed 5 August 2011, p. 21.
164. Ibid.
165. M. Sheehan (2007), 'China: The Long March into Space', *The International Politics of Space* (London: Routledge), p. 159.
166. Ibid.
167. Ibid., p. 170.
168. W. Rathgeber (2007), 'China's Posture in Space', *ESPI Report,* issue 3, June, http://www.espi.or.at/images/stories/dokumente/studies/espi_china_report_rev4-1_wf.pdf, date accessed 3 May 2011, p. 25.
169. M. S. Smith (2005), 'China's Space Program: An Overview', *CRS Report for Congress*, 15 October, http://www.fas.org/sgp/crs/space/RS21641.pdf, date accessed 3 May 2011.
170. D. E. Brock (2009), 'Science Innovation During the Cultural Revolution: Notes from the *Peking Review', Southeast Review of Asian Studies,* vol. 31, http://fordham.academia.edu/DarrylBrock/Papers/109783/Science_Innovation_during_the_Cultural_Revolution_Notes_from_the_Peking_Review, date accessed 5 August 2011, p. 226.
171. M. S. Smith (2003), 'China's Space Program: An Overview', *CRS Report for Congress*, 21 October, http://www.defenselink.mil/pubs/20030730chinaex.pdf, date accessed 3 May 2011.
172. Sheehan (2007), 'China: The Long March into Space', p. 161.
173. Y. Chen (1991), 'China's Space Policy: A Historical Review', *Space Policy*, vol. 7, issue 2, May, p. 128.
174. Y. Chen (2011), 'Dragon Week: Orbiting Dragons', 18 January, http://blog.heritage.org/2011/01/18/orbiting-dragons/, date accessed 3 May 2011.
175. Johnson-Freese (2007), 'China's Space Ambitions', p. 9.
176. Pagkratis (2010), 'Space Policies, Issues and Trends in 2009/2010', p. 52.
177. De Neve (2011), 'Reconsidering the Military Axis of the Chinese Space Programme', p. 4.

178. Johnson-Freese (2007), 'China's Space Ambitions', p. 11.
179. Information Office of the State Council of the People's Republic of China (2006), 'China's Space Activities in 2006', Section I-Aims and Principles of Development.
180. Pollpeter (2008), 'Building for the Future: China's Progress in Space Technology During the Tenth 5-Year Plan and the US Response', p. viii.
181. Johnson-Freese (2007), 'China's Space Ambitions', p. 5.
182. Cheng (2011), 'Dragon Week: Orbiting Dragons'.
183. Xinhua (2010), 'China Tests Space Station Module for 2011 Launch', 17 August, http://www.msnbc.msn.com/id/38740006/ns/technology_and_science-space/, date accessed 4 May 2011.
184. Xinhua (2011), 'China Sees Smooth Preparation for Launch of Unmanned Module,' 2 May, *Space Daily*, http://www.spacedaily.com/reports/China_sees_smooth_preparation_for_launch_of_unmanned_module_999.html, date accessed 4 May 2011.
185. J. Foust (2008), 'China and the US: Space Race or Miscommunication?' *Space Review*, 3 March, http://www.thespacereview.com/article/1075/1, date accessed 15 June 2011.
186. J. Foust (2003), 'China, Shenzhou, and the ISS', *Space Review*, 20 October, http://www.thespacereview.com/article/53/1, date accessed 15 June 2011.
187. 'What's the Future of US-China Cooperation in Space?' (2011), *Space Politics*.
188. D. Barboza (2008), 'Astronauts Return Safely to China', *New York Times*, 29 September, http://www.nytimes.com/2008/09/29/world/asia/29china.html?partner=rssnyt&emc=rss, date accessed 4 May 2011.
189. P. Besha (2010), 'Policy Making in China's Space Program: A History and Analysis of the Chang'e Lunar Orbiter Project', *Space Policy*, vol. 26, issue 4, November, p. 214.
190. J. Drew (2008), 'Space Inspires Passion and Practicality in China', *Washington Post*, 25 September, http://www.washingtonpost.com/wp-dyn/content/article/2008/09/23/AR2008092302649.html, date accessed 4 May 2011.
191. Johnson-Freese (2007), 'China's Space Ambitions', p. 13.
192. Ibid., p.14.
193. 'Aiming High: China is Moving Heaven and Earth to Put a Man on the Moon' (2009), *The Economist*, 22 October, http://www.economist.com/node/14678539?story_id=14678539, date accessed 8 February 2011.
194. Ibid.
195. W. Chunsi (2006), 'Development Goals of China's Space Program', *China Security*, issue 2, 21 March, http://www.wsichina.org/attach/cs2_9.pdf, date accessed 8 February 2011.
196. Sheehan (2007), 'China: The Long March into Space', p. 159.
197. Ibid., p. 160.
198. Ibid.
199. Johnson-Freese (2007), 'China's Space Ambitions', p. 17.
200. 'Space Security 2010', p. 78.
201. Acuthan (2006), 'China's Outer Space Programme: Diplomacy of Competition or Cooperation?', p. 4.
202. Ibid.
203. Ibid., p. 5.

204. F. Tronchetti (2011), 'Preventing the Weaponization of Outer Space: Is a Chinese-Russian-European Common Approach Possible?', *Space Policy*, vol. 27, issue 2, May, p. 84.
205. Sheehan (2007), 'China: The Long March into Space', p. 168.
206. Johnson-Freese (2007), 'China's Space Ambitions', p. 19.
207. B. W. MacDonald (2008), 'China, Space Weapons and US Security', *Council Special Report*, issue 38, September, http://www.cfr.org/content/publications/attachments/China_Space_CSR38.pdf, date accessed 15 June 2011, p. 4.
208. R. Weitz (2007), 'Special Report: Chinese Anti-Satellite Weapon Test, the Shot Heard Round the World', *WMD Insights*, March, http://www.wmdinsights.com/I13/I13_EA1_SP_II_RussiaChinese.htm, date accessed 28 March 2011, Part I: US Allies Criticize China's Anti-Satellite Weapon Test; Media Notes Concerns About US Space Policies.
209. G. Anzera (2007), 'China's ASAT Test and its Impact on the United States', *World Security Network*, 29 January, http://www.worldsecuritynetwork.com/showArticle3.cfm?article_id=13973, date accessed 5 August 2011.
210. T. Hitchens (2008), 'Space Wars', *Scientific American*, vol. 329, issue 3, March, p. 81.
211. 'Chinese ASAT Test' (2011), *Celes Trak*, 20 May, http://celestrak.com/events/asat.asp, date accessed 5 August 2011.
212. Weitz (2007), 'Special Report: Chinese Anti-Satellite Weapon Test, the Shot Heard Round the World', Part I: US Allies Criticize China's Anti-Satellite Weapon Test.
213. Ibid.
214. H. Zhang (2008), 'Chinese Perspectives on Space Weapons', in P. Podvig and H. Zhang, *Russian and Chinese Responses to US Military Plans in Space* (Cambridge: American Academy of Arts and Sciences, 2008), http://www.amacad.org/publications/militarySpace.pdf, date accessed 5 August 2011, p. 39.
215. P. Ritter (2008), 'The New Space Race: China vs. US', *Time Magazine*, 13 February, http://www.time.com/time/world/article/0,8599,1712812,00.html, date accessed 5 August 2011.
216. Cf. Acuthan (2006), 'China's Outer Space Programme: Diplomacy of Competition or Cooperation?'.
217. Pagkratis (2010), 'Space Policies, Issues and Trends in 2009/2010', p. 52.
218. Sheehan (2007), 'China: The Long March into Space', p. 169.
219. Ibid.
220. Ibid., p. 168.
221. Ibid., p. 162.
222. Ibid., p. 163.
223. Pollpeter (2008), 'Building for the Future: China's Progress in Space Technology During the Tenth 5-Year Plan and the US Response', p. viii.
224. Ibid., pp. 32–3.
225. Ibid.
226. Johnson-Freese (2007), 'China's Space Ambitions', p. 17.
227. Cf. Smith (2003), 'China's Space Program: An Overview'.
228. The China's 2009 space budget was reported to average USD 2.2 billion, as opposed to the US of about USD 39.5 billion. M. Krepon and S. Black (2009), *Space Security or Anti-Satellite Weapons?* (Washington, D.C.: The Henry L. Stimson Centre), p. 15.

229. J. Kynge (2006), *China Shakes the World: A Titan's Rise and Troubled Future – and the Challenge for America* (New York: Houghton Mifflin), p. 114, cited in Pollpeter (2008), 'Building for the Future: China's Progress in Space Technology During the Tenth 5-Year Plan and the US Response', p. 28.
230. Pollpeter (2008), 'Building for the Future', p. 31.
231. Ibid.
232. Ibid., pp. 2–3.
233. Ibid., p. 32.
234. Sheehan (2007), 'China: The Long March into Space', p. 167.
235. Pagkratis (2010), 'Space Policies, Issues and Trends in 2009/2010', p. 52.
236. 'Resetting US-China Space Cooperation' (2011), *Space Politics*, 20 January,http://www.spacepolitics.com/2011/01/20/resetting-us-china-space-cooperation/, date accessed 6 May 2011.
237. White House (2011), *US-China Joint Statement*, 19 January, Point 10, http://www.whitehouse.gov/the-press-office/2011/01/19/us-china-joint-statement, date accessed 6 May 2011.
238. Chunsi (2006), 'Development Goals of China's Space Program'.
239. Z. Jing (2008), 'China and Space Security', in J. M. Logsdon and J. C. Moltz (eds), *Collective Security in Space-Asian Perspectives* (Washington, D.C.: Space Policy Institute), http://www.gwu.edu/~spi/assets/docs/Collective%20Security%20in%20Space%20-%20Asian%20Perspectives%20-%20January%202008.pdf, date accessed 8 August 211, p. 80.
240. Johnson-Freese (2007), 'China's Space Ambitions', p. 26.
241. Ibid.
242. Rendleman and Faulconer (2010), 'Improving International Space Cooperation', p. 149.
243. Government of the Russian Federation (2005), 'Major Provisions of the Russian Federal Space Program for 2006–2015', *Federal Space Program of the Russian Federation for 2006–2015*, 22 October, http://www.infoespacial.com/wp-content/uploads/Major-provisions-of-the-Russian-Federal-Space-Program-for-2006-2015.pdf, date accessed 18 May 2005.
244. Ibid., Section: Goals of the Program.
245. Pagkratis (2010), 'Space Policies, Issues and Trends in 2009/2010', p. 49.
246. J. E. Oberg (2011), 'International Perspectives: Russia', in Lutes and Hays (eds), et al. (2011), *Toward a Theory of Spacepower*, p. 438.
247. Pagkratis (2010), 'Space Policies, Issues and Trends in 2009/2010', p. 49.
248. 'Space Security 2010', p. 114.
249. Government of the Russian Federation (2005), 'Major Provisions of the Russian Federal Space Program for 2006–2015', Section: Description of the Issue to be Solved by the Program.
250. A. Arbatov (2011), 'Russian Perspectives on Spacepower', in Lutes and Hays (eds), et al. (2011), *Toward a Theory of Spacepower*, p. 443.
251. Ibid., p. 443.
252. Rendleman and Faulconer (2010), 'Improving International Space Cooperation', pp. 144–5.
253. Oberg (2011), 'International Perspectives: Russia', in Lutes and Hays (eds), et al. (2011), *Toward a Theory of Spacepower*, p. 438.

254. Government of the Russian Federation (2005), 'Major Provisions of the Russian Federal Space Program for 2006–2015', Section: Description of the Issue to be Solved by the Program.
255. Ibid.
256. Arbatov (2011), 'Russian Perspectives on Spacepower', in Lutes and Hays (eds), et al. (2011), *Toward a Theory of Spacepower*, p. 444.
257. Ibid., p. 443.
258. Mathieu (2008), 'Assessing Russia's Space Cooperation with China and India', p. 16.
259. Government of the Russian Federation (2005), 'Major Provisions of the Russian Federal Space Program for 2006–2015', Section: Description of the Issue to be Solved by the Program.
260. Arbatov (2011), 'Russian Perspectives on Spacepower', in Lutes and Hays (eds), et al. (2011), *Toward a Theory of Spacepower*, p. 445.
261. Ibid.
262. 'Russia's New Security Strategy: Sleeker and Stronger' (2009), *TV-Novosti*, 13 May, http://rt.com/usa/news/russia-s-new-security-strategy-sleeker-and-stronger/, date accessed 18 May 2011.
263. 'Space Security 2010', p. 114.
264. Ibid.
265. Tronchetti (2011), 'Preventing the Weaponization of Outer Space: Is a Chinese-Russian-European Common Approach Possible?', p. 87.
266. Pagkratis (2010), 'Space Policies: Issues and Trends in 2009/2010', p. 50.
267. Arbatov (2011), 'Russian Perspectives on Spacepower' in Lutes and Hays (eds), et al. (2011), *Toward a Theory of Spacepower*, p. 444.
268. Ibid.
269. S. Garber (2007), 'Sputnik and the Dawn of the Space Age', NASA website, 10 October, http://history.nasa.gov/sputnik/, date accessed 18 May 2011.
270. Ibid.
271. A. J. LePage (1997), 'Sputnik II: The First Animal in Orbit', http://www.svengrahn.pp.se/histind/Sputnik2/SpaceViews%20November%201997%20Articles.htm, date accessed 5 August 2011.
272. 'Q&A: When Was the First Manned Mission up into Space?' *Space Today Online*, http://www.spacetoday.org/Questions/FirstMannedMission.html, date accessed 23 May 2011.
273. 'Yuri Gagarin: First Man in Space', NASA website, http://www.nasa.gov/mission_pages/shuttle/sts1/gagarin_anniversary.html, date accessed 23 May 2011.
274. P. Podvig (2008), 'Russia and Military Uses of Space', in P. Podvig and H. Zhang, *Russian and Chinese Responses to US Military Plans in Space* (Cambridge: American Academy of Arts and Sciences), http://www.amacad.org/publications/militarySpace.pdf, date accessed 23 May 2011, p. 5.
275. 'Yuri Gagarin: First Man in Space', NASA website.
276. Podvig (2008), 'Russia and Military Uses of Space', in Podvig and Zhang (2008), *Russian and Chinese Responses to US Military Plans in Space*, p. 5.
277. 'Yuri Gagarin: First Man in Space', NASA website.

278. A. Zak (2008), 'Russian Space Program: A Decade Review (2000–2010)', *Russia Space Web,* 3 January, http://www.russianspaceweb.com/russia_2000_2010.html, date accessed 23 May 2011.
279. Oberg (2011), 'International Perspectives: Russia', in Lutes and Hays (eds), et al. (2011), *Toward a Theory of Spacepower*, p. 425.
280. Government of the Russian Federation (2005), 'Major Provisions of the Russian Federal Space Program for 2006–2015', Section: Goal of the Program.
281. Ibid., Section: Description of the Issue to be Solved by the Program.
282. Pagkratis (2010), 'Space Policies, Issues and Trends in 2009/2010', p. 49.
283. Ibid.
284. G. Acar (2009), 'The Geopolitics of Satellite Navigation Technology: The Russian Alternative to GPS', *Eurasia Critic*, March, http://www.eurasiacritic.com/articles/geopolitics-satellite-navigation-technology-russian-alternative-gps, date accessed 19 May 2011.
285. Pagkratis (2010), 'Space Policies: Issues and Trends in 2009/2010', p. 49.
286. S. Pagkratis (2011), 'Space Policies: Issues and Trends in 2010/2011', *ESPI Report*, issue 35, June (Vienna: ESPI), p. 52.
287. Pagkratis (2010), 'Space Policies: Issues and Trends in 2009/2010', p. 50.
288. B. de Montluc (2010), 'Russia's Resurgence: Prospects for Space Policy and International Cooperation', *Space Policy*, vol. 26, issue 1, February, p. 16.
289. C. Moore (2000), 'Apollo-Soyuz Mission 25 Years Later', *ABC News*, 16 July, http://abcnews.go.com/Technology/story?id=120107&page=1&page=1, date accessed 23 May 2011.
290. V. Putkov (2007), 'Sputnik and Russia's Outer Space Activities', in UNIDIR, 'Celebrating the Space Age: 50 Years of Space Technology, 40 Years of the Outer Space Treaty', Conference Report 2–3 April (New York and Geneva: United Nations), http://www.unidir.org/pdf/articles/pdf-art2661.pdf, date accessed 8 August 2011, p. 39.
291. Moore (2000), 'Apollo-Soyuz Mission 25 Years Later'.
292. J. I. Petty (2004), 'Spacecraft: Mir Space Station', *NASA* website, 4 April, http://spaceflight.nasa.gov/history/shuttle-mir/spacecraft/s-mir.htm, date accessed 23 May 2011.
293. D. Whitehouse (2001), 'Mir: A Home in Space', *BBC News*, 23 March, http://news.bbc.co.uk/2/hi/science/nature/1225005.stm, date accessed 23 May 2011.
294. 'Mir Space Station 1986–2001', *BBC News*, http://news.bbc.co.uk/hi/english/static/in_depth/sci_tech/2001/mir/default.stm, date accessed 23 May 2011.
295. Petty (2004), 'Spacecraft: Mir Space Station'.
296. 'Mir Space Station 1986–2001'.
297. M. Sheehan (2007), The International Politics of Space (New York and London: Routledge), p. 179.
298. J. M. Logsdon and J. R. Millar (2001), 'US-Russian Cooperation in Human Space flight: Assessing the Impacts', *Space Policy*, vol. 17, issue 3, August, p. 171.
299. 'The International Space Station', *Shuttlepresskit.com*, http://www.shuttlepresskit.com/ISS_OVR/index.htm, date accessed 23 May 2011.
300. Government of the Russian Federation (2005), 'Major Provisions of the Russian Federal Space Program for 2006–2015', Section: Description of the Issues to be Resolved by the Program.

301. A. Kizilova (2009), 'Russian Priorities in Space', *Russia-IC.com*, 27 January, http://www.russia-ic.com/education_science/science/science_overview/862/, date accessed 23 May 2011.
302. Government of the Russian Federation (2005), 'Major Provisions of the Russian Federal Space Program for 2006–2015', Section: Description of the Issues to be Resolved by the Program.
303. Oberg (2011), 'International Perspectives: Russia', in Lutes and Hays (eds), et al. (2011), *Toward a Theory of Spacepower*, p. 434.
304. Ibid.
305. Mathieu (2008), 'Assessing Russia's Space Cooperation with China and India', p. 5.
306. S. Lavrov cited in 'Russia Proposes Treaty to Ban Space Weapons' (2008), *New Scientist*, 12 February, http://space.newscientist.com/article/dn13312-russia-proposes-treaty-to-ban-space-weapons.html, date accessed 23 May 2011.
307. Statement by S. Lavrov at the plenary meeting of the Conference on Disarmament (2008), Geneva, 12 February, http://www.reachingcriticalwill.org/political/cd/speeches08/1session/Feb12RussiaLavrov.pdf, date accessed 8 August 2011, p. 6.
308. T. Hitchens (2010), 'Multilateralism in Space: Opportunities and Challenges for Achieving Space Security', Space and Defense, vol. 4, issue 2, Summer, http://www.unidir.ch/unidir-views/pdf/pdf-uv-29-32.pdf, date accessed 15 August 2011, p. 19.
309. United Nations General Assembly (2010), 'Transparency and Confidence-Building Measures in Outer Space Activities', Sixty-Fifth Session of the First Committee, Agenda Item 97 (y), General and Complete Disarmament: Transparency and Confidence-Building Measures in Outer Space Activities, A/C.1/65/L.38, 15 October, http://daccess-dds-ny.un.org/doc/UNDOC/LTD/N10/586/81/PDF/N1058681.pdf?OpenElement, date accessed 19 August 2011.
310. Ibid.
311. Ibid.
312. United Nations General Assembly (2011), 'Transparency and Confidence-Building Measures in Outer Space Activities', A/RES/65/68, 13 January, http://www.reachingcriticalwill.org/political/1com/1com10/gares/A%20RES%2065%2068.pdf, date accessed 19 August 2011.
313. Tronchetti (2011), 'Preventing the Weaponization of Outer Space: Is a Chinese-Russian-European Common Approach Possible?', p. 87.
314. Government of the Russian Federation (2005), 'Major Provisions of the Russian Federal Space Program for 2006–2015', Section: Principal Objectives of the Program.
315. 'Space Security 2010', p. 89.
316. Government of the Russian Federation (2005), 'Major Provisions of the Russian Federal Space Program for 2006–2015', Section: Description of the Issues to be Resolved by the Program.
317. Ibid.
318. Ibid.
319. Ibid.
320. de Montluc (2010), 'Russia's Resurgence: Prospects for Space Policy and International Cooperation', p. 22.

321. Arbatov (2011), 'Russian Perspectives on Spacepower', in Lutes and Hays (eds), et al. (2011), *Toward a Theory of Spacepower*, p. 444.
322. 'Space Security 2010', p. 124.
323. Arbatov (2011), 'Russian Perspectives on Spacepower', in Lutes and Hays (eds), et al. (2011), *Toward a Theory of Spacepower*, p. 445.
324. Ibid.
325. D. Wolter (2006), *Common Security in Outer Space and International Law* (Geneva: United Nations UNIDIR), http://www.unidir.org/pdf/ouvrages/pdf-1-92-9045-177-7-en.pdf, date accessed 8 August 2011, p. 37.
326. Ibid., p. 38.
327. Ibid., p. 22.
328. Zak (2008), 'Russian Space Program: A Decade Review (2000–2010)'.
329. Podvig (2008), 'Russia and Military Uses of Space', in Podvig and Zhang (2008), *Russian and Chinese Responses to US Military Plans in Space*, p. 2.
330. Ibid., pp. 1–2.
331. L. Grego (2003), 'Short History of US and Soviet ASAT Programs', Union of Concerned Scientists, 1 April, http://www.ucsusa.org/assets/documents/nwgs/asat_history.pdf, date accessed 8 August 2011, p. 3.
332. Kizilova (2009), 'Russian Priorities in Space'.
333. Pagkratis (2010), 'Space Policies: Issues and Trends in 2009/2010', p. 50.
334. Ibid.
335. 'Space Security 2010', p. 129.
336. Pagkratis (2010), 'Space Policies: Issues and Trends in 2009/2010', p. 50.
337. Oberg (2011), 'International Perspectives: Russia', in Lutes and Hays (eds), et al. (2011), *Toward a Theory of Spacepower*, p. 425.
338. C. Lindborg (1997), 'Baikonur Cosmodrome', Federation of American Scientists (FAS) website, http://www.fas.org/spp/guide/russia/facility/baikonur.htm, date accessed 23 May 2011.
339. Podvig (2008), 'Russia and Military Uses of Space', in Podvig and Zhang (2008), *Russian and Chinese Responses to US Military Plans in Space*, p. 5.
340. Pagkratis (2010), 'Space Policies: Issues and Trends in 2009/2010', p. 49.
341. S. Radosevic (2003), 'Patterns of Preservation, Restructuring and Survival: Science and Technology Policy in Russia in Post-Soviet Era', *Research Policy*, vol. 32, pp. 1105–1124.
342. Ibid., p. 1106.
343. Mathieu (2008), 'Assessing Russia's Space Cooperation with China and India', p. 14.
344. Ibid.
345. Ibid., p. 16.
346. Government of the Russian Federation (2005), 'Major Provisions of the Russian Federal Space Program for 2006–2015', Section: Description of the Issues to be Resolved by the Program.
347. Kizilova (2009), 'Russian Priorities in Space'.
348. Ibid.
349. Podvig (2008), 'Russia and Military Uses of Space', in Podvig and Zhang (2008), *Russian and Chinese Responses to US Military Plans in Space*, p. 1.
350. Key components of this restructuring are the division between civil and military, and between the aeronautics and space programmes. In addition, industrial reorganizations have taken place, and the Russian space agency,

Roskosmos, is now exclusively focused on civil space. de Montluc (2010), 'Russia's Resurgence: Prospects for Space Policy and Internal Cooperation', pp. 15–16.

351. de Montluc (2010), 'Russia's Resurgence: Prospects for Space Policy and Internal Cooperation', pp. 15–16.
352. Mathieu (2008), 'Assessing Russia's Space Cooperation with China and India', p. 17.
353. Ibid., p. 13.
354. Ibid., pp. 15–16.
355. H. R. Hertzfeld (2011), 'Commercial Space and Spacepower', in Lutes and Hays (eds), et al. (2011), *Toward a Theory of Spacepower*, p. 83.
356. Ibid.
357. 'Company Profile', Key Documents, EADS website, http://www.eads.com/eads/int/en/news/key-Documents/About-EADS.html, date accessed 14 July 2011.
358. 'The Company: What We Do and Who We Are', EADS-Astrium website, http://www.astrium.eads.net/en/who-is-astrium/, date accessed 14 July 2011.
359. 'Activities', EADS-Astrium website, http://www.astrium.eads.net/en/our-expertises/, date accessed 14 July 2011.
360. 'Galileo. The European Satellite Navigation System', EADS-Astrium website, http://www.astrium.eads.net/en/programme/galileo.html, date accessed 18 July 2011.
361. 'Ariane 5: Placing the Heaviest Loads in Orbit', EADS-Astrium website, http://www.astrium.eads.net/en/programme/ariane-5.html, date accessed 18 July 2011.
362. 'ATV (Automated Transfer Vehicle) 1 & 2', EADS-Astrium website, http://www.astrium.eads.net/en/programme/atv.html, date accessed 18 July 2011.
363. Ibid.
364. 'Flying Colours for the ATV Johannes Kepler Mission' (2011), *Astrium News*, 24 June, http://www.astrium.eads.net/en/news2/flying-colours-for-the-atv-johannes-kepler-mission.html, date accessed 19 July 2011.
365. Ibid.
366. 'Activities', EADS-Astrium website, http://www.astrium.eads.net/en/our-expertises/, date accessed 14 July 2011.
367. Ibid.
368. 'SPOT Maps: High-resolution Colour Satellite Images', EADS-Astrium website, http://www.spotimage.com/web/en/1285-spotmaps.php, date accessed 15 July 2011.
369. 'Telecom Services', EADS-Astrium website, http://www.astrium.eads.net/en/telecom-services/, date accessed 15 July 2011.
370. 'From Stabilisation to Expansion: EADS Reports 2010 Full Year Results', EADS website, 9 March, http://www.eads.com/eads/int/en/news/press.379c9037-0dc0-4fea-a125-88a94cde84c0.70472f39-dd6f-4428-a792-91d82cb9791b.html, date accessed 15 July 2011.
371. The space programmes discussed in the grid are only a few examples, selected to illustrate the activities of Astrium in a particular sphere and are

not an exhaustive list of all the programmes it is involved in. For more details, see http://www.astrium.eads.net/en/programme/, date accessed 15 July 2011.

372. 'GMES. Global Monitoring for Environment and Security: The Smart Way of Looking at the World', EADS-Astrium website, http://www.astrium.eads.net/en/programme/gmes.html, date accessed 15 July 2011.
373. 'EarthCARE', EADS-Astrium website, http://www.astrium.eads.net/en/programme/earthcare.html, date accessed 15 July 2011.
374. 'ERSSS. Partnership for Kazakhstan's Space Programme', EADS-Astrium website, http://www.astrium.eads.net/en/programme/ersss.html, date accessed 15 July 2011.
375. 'Services. Geo-Information Services', EADS-Astrium website, http://www.astrium.eads.net/en/geo-information-services/, date accessed 15 July 2011.
376. 'The Company: What We Do and Who We Are', EADS-Astrium website.
377. 'TerraSAR-X. The Highly Capable Synthetic Aperture Radar Satellite', EADS-Astrium website, http://www.astrium.eads.net/en/programme/terrasar-x.html, date accessed 15 July 2011.
378. 'Columbus: The European Space Laboratory', EADS-Astrium website, http://www.astrium.eads.net/en/programme/columbus.html, date accessed 17 July 2011.
379. Ibid.
380. 'Telecom Services. Our Products', EADS-Astrium website, http://www.astrium.eads.net/en/telecom-services/, date accessed 15 July 2011.
381. Ibid.
382. EADS-Astrium website, http://www.infoterra-global.com/applications_defense, date accessed 17 July 2011.
383. 'Boeing in Brief', Boeing website, http://www.boeing.com/companyoffices/aboutus/brief.html, date accessed 20 July 2011.
384. Ibid.
385. 'Boeing: Defense, Space and Security', Boeing website, http://www.boeing.com/bds/overview.html, date accessed 20 July 2011.
386. Ibid.
387. Ibid.
388. Boeing (2011), 'Boeing Network and Space Systems: Connecting Today with Tomorrow', Boeing Brochure, http://www.boeing.com/bds/network_space/files/NandSS_brochure.pdf, date accessed 20 July 2011, p. 7.
389. 'Boeing: Defense, Space and Security', Boeing website.
390. Ibid.
391. Boeing (2011), 'Boeing Network and Space Systems: Connecting Today with Tomorrow', p. 10.
392. 'Boeing: Defense, Space and Security', Boeing website.
393. 'Boeing Space Exploration', Boeing website, http://www.boeing.com/defense-space/space_exploration/backgrounder.pdf, date accessed 20 July 2011, p. 2.
394. Ibid.
395. 'Global Positioning System GPS IIF', Boeing website, http://www.boeing.com/defense-space/space/gps/, date accessed 20 July 2011.

396. 'Boeing: Defense, Space and Security', Boeing website.
397. Z. Meyer (2010), 'Private Commercialization of Space in an International Regime: A Proposal for a Space District', *Northwestern Journal of International Law and Business*, vol. 30, issue 1, Winter, http://www.iew.unibe.ch/unibe/rechtswissenschaft/dwr/iew/content/e3870/e3985/e4139/e6404/sel-topic_4-privatecommercial_ger.pdf, date accessed 24 May 2011, p. 246.
398. Space Foundation (2011), 'The Authoritative Guide to Global Space Activity', p. 4.
399. Skaar (2007), 'Commercialisation of Space and Its Evolution', p. 5.
400. Ibid., p. 13.
401. Meyer (2010), 'Private Commercialization of Space in an International Regime', pp. 246–247.
402. 'Space Security 2010', p. 103.
403. Ibid., p. 9.
404. Ibid.
405. B. Warf (2007), 'Geopolitics of the Satellite Industry', *Tijdschrift voor Economische en Sociale Geografie*, vol. 98, issue 3, p. 385.
406. 'Space Security 2010', p. 109.
407. Ibid.
408. Skaar (2007), 'Commercialisation of Space and its Evolution', p. 16.
409. J. A. Giacalone (2008), 'Global Trends in the Commercialization of Space', *Journal of Business & Economics Research*, vol. 6, issue 8, August, http://journals.cluteonline.com/index.php/JBER/article/view/2462/2508, date accessed 8 August 2011, pp. 68–9.
410. Homans (2010), 'The Wealth of Constellations'.
411. 'Space Security 2010', p. 102.
412. Skaar (2007), 'Commercialisation of Space and its Evolution', p. 27.
413. 'Space Security 2010', p. 102.
414. Homans (2010), 'The Wealth of Constellations'.
415. Ibid.
416. Ibid.
417. 'Space Security 2010', p. 115.
418. Ibid., p. 116.
419. Ibid., p. 22.
420. Ibid., p. 103.
421. Pagkratis (2010), 'Space Policies: Issues and Trends in 2009/2010', p. 87.
422. 'Space Security 2010', p. 107.
423. Pagkratis (2011), 'Space Policies: Issues and Trends in 20010/2011', p. 16.
424. J. Fuller, J. Foust, C. Frappier, D. Kaiser and D. Vaccaro (2011), 'The Commercial Space Industry: A Critical Spacepower Consideration', in Lutes and Hays (eds), et al. (2011), *Toward a Theory of Spacepower*, p. 107.
425. Ibid., p. 111.
426. Ibid.
427. Ibid.
428. Ibid.
429. Pagkratis (2010), 'Space Policies: Issues and Trends in 2009/2010', p. 87.
430. 'Space Security 2010', p. 107.

431. Ibid., p. 108.
432. Fuller et al. (2011), 'The Commercial Space Industry: A Critical Spacepower Consideration', in Lutes and Hays (eds), et al. (2011), *Toward a Theory of Spacepower*, p. 107.
433. Pagkratis (2011), 'Space Policies: Issues and Trends in 20010/2011', p. 20.
434. Ibid., pp. 26–7.
435. Ibid., p. 26.
436. Ibid.
437. 'Space Security 2010', p. 108.
438. Center for Defense Information (2008), 'Space Security Update 3', CDI, 21 May, http://www.cdi.org/program/issue/document.cfm?DocumentID=4298&IssueID=140&StartRow=1&ListRows=10&appendURL=&Orderby=DateLastUpdated&ProgramID=68&issueID=140, date accessed 1 June 2011.
439. Space Foundation (2011), 'The Authoritative Guide to Global Space Activity', p. 6.
440. Ibid.
441. Fuller et al. (2011), 'The Commercial Space Industry: A Critical Spacepower Consideration', in Lutes and Hays (eds), et al. (2011) *Toward a Theory of Spacepower*, p. 118.
442. Space Foundation (2011), 'The Authoritative Guide to Global Space Activity', p. 6.
443. Fuller et al. (2011), 'The Commercial Space Industry: A Critical Spacepower Consideration', in Lutes and Hays (eds), et al. (2011) *Toward a Theory of Spacepower*, p. 123'.
444. Meyer (2010), 'Private Commercialization of Space in an International Regime', pp. 243–244.
445. Ibid., p. 245.
446. L. Nardon and C. Venet (2011), 'The Development of Public-Private Partnerships in the European Satcom Sector', The Europe & Space Series, issues 4, May, <http://www.ifri.org/?page=detail-contribution&id=6607>, date accessed 8 August 2011, p. 4.
447. Meyer (2010), 'Private Commercialization of Space in an International Regime', p. 247.
448. Ibid.
449. 'Space Security 2010', p. 19.
450. Space Foundation (2011), 'The Authoritative Guide to Global Space Activity', p. 4.
451. Ibid.
452. Ibid.
453. Ibid., p. 4.
454. N. Peter (2007), 'Space Policy: Issues and Trends in 2006/2007', *ESPI Report*, issue 6, September, http://www.espi.or.at/images/stories/dokumente/studies/6th%20espi%20report.pdf, date accessed 8 August 2011, p. 62.
455. Ibid.
456. 'Space Security 2010', p. 107.
457. Pagkratis (2010), 'Space Policies, Issues and Trends in 2009/2010', p. 114.
458. Hertzfeld (2011), 'Commercial Space and Spacepower', in Lutes and Hays (eds), et al. (2011) *Toward a Theory of Spacepower*, p. 83.
459. United Nations General Assembly (1966), 'Treaty on Principles Governing the Activities of States in the Exploration and Use of Outer Space, including the Moon and Other Celestial Bodies', Res. 2222 (XXI), 19 December, Preamble.

6 Space Governance and *Meta*-Geopolitics

1. Space Foundation (2008), *The Space Report 2008: The Authoritative Guide to Global Space Activity*, Executive Summary, http://www.contentfirst.com/past/Spacefoundation/08executivesummary.pdf, date accessed 2 August 2011, p. 8.
2. T. S. Miller (2004), 'Partnership: The Way of the Future for The International Space Station', *NPMA*, vol. 16, issue 5, http://www.npma.org/Archives/vol.16-5-Miller.pdf, date accessed 8 August 2011, p. 8.
3. E. Sadeh (2004), 'Technical, Organizational and Political Dynamics of the International Space Station Program', *Space Policy*, vol. 20, issue 3, August, p. 186.
4. 'International Space Station: International Cooperation', NASA website, http://www.nasa.gov/mission_pages/station/cooperation/index.html, date accessed 23 July 2011.
5. Commission of the European Communities (2007), 'European Space Policy', Communication from the Commission to the Council and the European Parliament, COM (2007) 212 final, 26 April, Brussels, http://eur-lex.europa.eu/LexUriServ/LexUriServ.do?uri=COM:2007:0212:FIN:en:PDF, date accessed 5 August 2011, p. 9.
6. J. R. Minkel (2010) 'Is the International Space Station Worth $100 Billion?' *Space.com*, 1 November, http://www.space.com/9435-international-space-station-worth-100-billion.html, date accessed 13 July 2011.
7. 'About International Space Station', European Space Agency (ESA) website, http://www.esa.int/esaHS/ESA6NE0VMOC_iss_0.html, date accessed 13 July 2011.
8. Women's International League for Peace and Freedom (WILPF), 'Outer Space: Background and History', in *Reaching Critical Will*, http://www.reachingcriticalwill.org/legal/paros/osbackground.html, date accessed 13 July 2011.
9. 'Functions and Powers of the General Assembly', United Nations website, http://www.un.org/en/ga/about/background.shtml, date accessed 13 July 2011.
10. United Nations General Assembly (1957), 'Regulation, Limitation and Balanced Reduction of All Armed Forces and All Armaments: Conclusion of an International Convention (treaty) on the Reduction of Armaments and the Prohibition of Atomic, Hydrogen and Other Weapons of Mass Destruction', Res. 1148 (XII), 14 November, http://daccess-dds-ny.un.org/doc/RESOLUTION/GEN/NR0/119/31/IMG/NR011931.pdf?OpenElement, date accessed 8 August 2011.
11. United Nations General Assembly (1961), 'International Co-operation in the Peaceful Uses of Outer Space', Res. 1721 (XVI), 20 December, http://www.oosa.unvienna.org/oosa/SpaceLaw/gares/html/gares_16_1721.html, date accessed 8 August 2011, quoted in D. Wolter (2006), *Common Security in Outer Space and International Law* (Geneva: United Nations UNIDIR), http://www.unidir.org/pdf/ouvrages/pdf-1-92-9045-177-7-en.pdf, date accessed 8 August 2011, p. 12.
12. United Nations General Assembly (1963), 'Declaration of Legal Principles Governing the Activities of States in the Exploration and Use of Outer Space', Res. 1962 (XVII), 13 December, http://www.oosa.unvienna.org/oosa/SpaceLaw/lpos.html, date accessed 23 July 2011.
13. United Nations General Assembly (1981), 'General and Complete Disarmament: A Study on Conventional Disarmament', A/RES/36/97,

9 December, http://www.un.org/documents/ga/res/36/a36r097.htm, date accessed 20 August 2011.

14. T. Hitchens (2010), 'Multilateralism in Space: Opportunities and Challenges for Achieving Space Security', *Space and Defense*, vol. 4, issue 2, summer, http://www.unidir.ch/unidir-views/pdf/pdf-uv-29-32.pdf, date accessed 15 August 2011, pp. 9–10.
15. G. Brachet (2006) 'Leveraging the Existing UN Space Machinery for Sustainable and Secure Access to Outer Space', in UNIDIR, 'Building the Architecture for Sustainable Space Security', Conference Report, 30–1 March (New York and Geneva: United Nations), p. 107.
16. United Nations General Assembly (1959), 'International Cooperation and the Peaceful Uses of Outer Space', Res. 1472 (XIV), 12 December, http://www.oosa.unvienna.org/oosa/en/SpaceLaw/gares/html/gares_14_1472.html, date accessed 13 July 2011.
17. 'United Nations Committee on the Peaceful Uses of Outer Space', UNOOSA website, http://www.oosa.unvienna.org/oosa/COPUOS/copuos.html, date accessed 8 August 2011.
18. Hitchens (2010), 'Multilateralism in Space', p. 15.
19. Brachet (2006) 'Leveraging the Existing UN Space Machinery for Sustainable and Secure Access to Outer Space', p. 107.
20. Hitchens (2010), 'Multilateralism in Space', p. 5.
21. United Nations General Assembly (1966), 'Treaty on Principles Governing the Activities of States in the Exploration and Use of Outer Space, including the Moon and Other Celestial Bodies', Res. 2222 (XXI), adopted on 19 December 1966, opened for signature on 27 January 1967, entered into force on 10 October 1967 (hereinafter 'Outer Space Treaty'), http://www.oosa.unvienna.org/oosa/en/SpaceLaw/gares/html/gares_21_2222.html, date accessed 21 July 2011.
22. United Nations, General Assembly (1967), 'Agreement on the Rescue of Astronauts, the Return of Astronauts and the Return of Objects Launched into Outer Space', Res. 2345 (XXII), adopted on 19 December 1967, opened for signature on 22 April 1968, entered into force on 3 December 1968, http://www.oosa.unvienna.org/oosa/SpaceLaw/gares/html/gares_22_2345.html, date accessed 21 July 2011.
23. United Nations General Assembly (1971), 'Convention on International Liability for Damage Caused by Space Objects', Res. 2777 (XXVI), adopted on 29 November 1971, opened for signature on 29 March 1972, entered into force on 1 September 1972, http://www.oosa.unvienna.org/oosa/SpaceLaw/gares/html/gares_26_2777.html, date accessed 21 July 2011.
24. United Nations General Assembly (1974), 'Convention on Registration of Objects Launched into Outer Space', Res. 3235 (XXIX), adopted on 12 November 1974, opened for signature on 14 January 1975, entered into force on 15 September 1976, http://www.unoosa.org/oosa/SORegister/regist.html, date accessed 21 July 2011.
25. United Nations General Assembly (1979), 'Agreement Governing the Activities of States on the Moon and Other Celestial Bodies', Res. 34/68, adopted on 5 December 1979, opened for signature on 18 December 1979, entered into force on 11 July 1984, http://www.oosa.unvienna.org/oosa/SpaceLaw/gares/html/gares_34_0068.html, date accessed 21 July 2011.

26. COPUOS Scientific and Technical Subcommittee (2007), 'Report of the Scientific and Technical Subcommittee of the Committee on Peaceful Uses of Outer Space on its Forty-fourth Session Held in Vienna from 12–23 February 2007', A/AC.105/890, 6 March, http://www.unoosa.org/pdf/reports/ac105/AC105_890E.pdf, date accessed 9 August 2011, p. 43.
27. United Nations General Assembly (2008), 'International Cooperation in the Peaceful Uses of Outer Space', A/RES/62/217, 1 February, http://www.oosa.unvienna.org/pdf/gares/ARES_62_217E.pdf, date accessed 9 August 2011.
28. COPUOS Scientific and Technical Subcommittee (2007), 'Report of the Scientific and Technical Subcommittee of the Committee on Peaceful Uses of Outer Space on its Forty-fourth Session Held in Vienna from 12–23 February 2007', p. 43.
29. Ibid., pp. 43–5.
30. COPUOS Legal Subcommittee (2007), 'Practice of States and International Organizations in Registering Space Objects', Working Paper submitted by the Chairman of the Working Group on the Practice of States and International Organizations in Registering Space Objects, A/AC.105/C.2/L.266, 30 January, http://www.oosa.unvienna.org/pdf/limited/c2/AC105_C2_L266E.pdf, date accessed 9 August 2011.
31. *Space Security 2008*, Spacesecurity.org, August, http://www.spacesecurity.org/SSI2008.pdf, date accessed 3 August 2011, p. 59.
32. T. Hitchens (2008), 'COPUOS Wades into the Next Great Space Debate', *Bulletin of the Atomic Scientists*, 26 June, http://www.thebulletin.org/web-edition/features/copuos-wades-the-next-great-space-debate, date accessed 9 August 2011.
33. *Space Security 2008*, p. 58.
34. COPUOS (2007), 'Future Role and Activities of the Committee on the Peaceful Uses of Outer Space', Working paper submitted by the Chairman, A/AC.105/L.268, 10 May, http://www.oosa.unvienna.org/pdf/limited/l/AC105_L268E.pdf, date accessed 9 August 2011.
35. Ibid.
36. Hitchens (2008), 'COPUOS Wades into the Next Great Space Debate'.
37. Cf. G. Brachet (2009), 'Best Practices for Space: A Necessary Underpinning for the Sustainability of Space Activities', UNIDIR Space Security Conference in Geneva 15–16 June, http://www.unidir.ch/pdf/conferences/pdf-conf115.pdf, date accessed 9 August 2011.
38. A. Bander (2010), 'COPUOS and the Long-Term Sustainability of Space Activities', Center for Strategic and International Studies (CSIS) website, 17 June, http://csis.org/blog/copuos-and-long-term-sustainability-space-activities, date accessed 23 July 2011.
39. The UN Office for Outer Space Affairs (UNOOSA) website, http://www.unoosa.org/oosa/en/OOSA/index.html, date accessed 9 August 2011.
40. 'UN Office for Outer Space Affairs. History of the Programme', UNOOSA website, http://www.unoosa.org/oosa/en/SAP/history.html, date accessed 9 August 2011.
41. 'United Nations Programme on Space Applications', UNOOSA website, <http://www.oosa.unvienna.org/oosa/en/about_PSA.html>, date accessed 9 August 2011.
42. 'UN Office for Outer Space Affairs. History of the Programme'.

43. C. A. Yepes (2009), 'Towards a UN Space Policy', *ESPI Perspectives,* issue 23, June, p. 6.
44. 'Disarmament in Geneva', United Nations Office at Geneva (UNOG) website, http://www.unog.ch/80256EE600585943/(httpHomepages)/6A03113D1857348E80256F04006755F6?OpenDocument, date accessed 13 July 2011.
45. *Space Security 2010*, Spacesecurity.org, August, http://www.spacesecurity.org/space.security.2010.reduced.pdf, date accessed 20 April 2011, p. 70.
46. Ibid.
47. Hitchens (2010), 'Multilateralism in Space', p. 15.
48. F. Tronchetti (2011), 'Preventing the Weaponization of Outer Space: Is a Chinese-Russian-European Common Approach Possible?', *Space Policy*, vol. 27, issue 2, May, p. 83.
49. United Nations, Conference on Disarmament (2009), 'Decision for the Establishment of a Programme of Work for the 2009 Session', CD/1864, 29 May, http://www.unog.ch/80256EDD006B8954/(httpAssets)/E8846993B5213D59C12575DF0029EE11/$file/CD+1864+English.pdf, date accessed 9 August 2011, p. 1, Point 2.
50. Ibid., p. 2, Point 3.
51. 'Protecting Free Access to Space' (2006), Remarks to the United Nations General Assembly First Committee, 11 October, in Women's International League for Peace and Freedom (WILPF), *Reaching Critical Will*, http://www.reachingcriticalwill.org/political/1com/1com06/statements/USoct11.pdf, date accessed 9 August 2011, p. 1.
52. M. Sethi (2011), 'A North Korean Embarrassment', *The Diplomat*, 22 July, http://the-diplomat.com/flashpoints-blog/2011/07/22/a-north-korean-embarrassment/, date accessed 23 July 2011.
53. Hitchens (2010), 'Multilateralism in Space', p. 17.
54. Tronchetti (2011), 'Preventing the Weaponization of Outer Space', p. 84.
55. Ibid.
56. B. S. Klapper (2008), 'Russia, China Challenge US Space Arms', *USA Today,* 2 December, http://www.usatoday.com/news/world/2008-02-12-2744020758_x.htm, date accessed 9 August 2011.
57. Hitchens (2010), 'Multilateralism in Space', p. 18.
58. Commission of the European Communities (2007), 'European Space Policy', p. 9.
59. 'Executive Summary', in *Space Security 2007*, Spacesecurity.org, August, http://www.spacesecurity.org/SSI2007ExecutiveSummary.pdf, date accessed 29 March 2011, p. 12.
60. Group on Earth Observations website, http://www.earthobservations.org, date accessed 9 August 2011.
61. 'About Us. Participating Organizations', Group on Earth Observations website, http://www.earthobservations.org/ag_partorg.shtml, date accessed 23 July 2011.
62. 'International Committee on Global Navigation Satellite Systems (ICG)', OOSA website, http://www.unoosa.org/oosa/SAP/gnss/icg.html, date accessed 9 August 2011.
63. International Committee on Global Navigation Satellite Systems (2007), 'A Forum to Discuss Global Navigation Satellite Systems (GNSS) to Benefit People Around the World', Brochure (Vienna: Secretariat of the International Committee on Global Navigation Satellite Systems),

http://www.oosa.unvienna.org/pdf/publications/icg_book01E.pdf, date accessed 9 August 2011.

64. UNOOSA (2008), 'Disaster Management and Space Technology: Bridging the Gap', Second United Nations International UN-SPIDER Workshop, Bonn, 13–15 October, http://earthobservations.org/meetings/20081013_15_2nd_un_spider_1st_announcement.pdf, date accessed 9 August 2011.
65. 'About UN-Spider', UNOOSA website, http://www.oosa.unvienna.org/oosa/en/unspider/index.html, date accessed 9 August 2011.
66. United Nations General Assembly (2007), 'Recent Launch of United Nations Platform for Space-Based Information for Disaster Management, Emergency Response (SPIDER) Hailed in Fourth Committee', GA/SPD/380, 25 October, http://www.un.org/News/Press/docs/2007/gaspd380.doc.htm, date accessed 9 August 2011.
67. Cf. R. Backhaus, N. Rohner and K.-U. Schrogl (2008), 'Satellites and Services for the Society: Visions and Perspectives', *Proceedings of the International Week on Space Applications*, Toulouse, http://www.espi.or.at/images/stories/dokumente/presentations/2008/paper_rohner_backhaus_schrogl_spaceappli08_2col.pdf, date accessed 9 August 2011.
68. Ibid.
69. Ibid.
70. 'What is GEOSS?: The Global Earth Observation System of Systems', Group on Earth Observations website, http://earthobservations.org/geoss.shtml, date accessed 9 August 2011.
71. United Nations (2004), *World Population to 2300* (New York: United Nations Publications), http://www.un.org/esa/population/publications/longrange2/WorldPop2300final.pdf, date accessed 9 August 2011, p. 4.
72. M. J. Peterson (1997), 'The Use of Analogies in Developing Outer Space Law', *International Organizations*, vol. 55, issue 1, Spring, p. 252.
73. Ibid.
74. Ibid., p. 253.
75. Ibid., p. 254.
76. 'Bogotá Declaration' (1976), *Declaration of the First Meeting of Equatorial Countries*, 3 December, http://www.spacelaw.olemiss.edu/library/space/International_Agreements/declarations/1976_bogota_declaration.pdf, date accessed 9 August 2011. The Declaration was signed by Brazil, Colombia, Congo, Ecuador, Indonesia, Kenya, Uganda and Zaire.
77. N. Jasentuliyana (1999), *International Space Law and the United Nations* (The Hague: Kluwer Law International), p. 153.
78. United Nations General Assembly (1962), 'Declaration of Legal Principles Governing the Activities of States in the Exploration and Use of Outer Space', Res. 1962 (XVII), adopted on 13 December 1963, http://www.oosa.unvienna.org/oosa/SpaceLaw/lpos.html, date accessed 23 July 2011.
79. A. Lambremont Webre (2005), 'Directions Towards an Exopolitics Initiative', *World Affairs, The Journal of International Issues*, vol. 9, issue 4, Winter, http://jerrypippin.com/WA-EXOPOLITICS%20INITIATIVE.pdf, date accessed 28 August 2011, p. 2.
80. W. Burr and H. L. Montford (2003), 'The Making of the Limited Test Ban Treaty, 1958–1963', *National Security Archive*, 8 August, http://www.gwu.edu/~nsarchiv/NSAEBB/NSAEBB94, date accessed 9 August 2011.

81. 'Outer Space Treaty' (1967) in United Nations (2002), *United Nations Treaties and Principles on Outer Space* (New York: United Nations), http://www.oosa.unvienna.org/pdf/publications/STSPACE11E.pdf, date accessed 10 August 2011, pp. 3–8.
82. 'Treaty Signatures', UNOOSA website, http://www.oosa.unvienna.org/oosatdb/showTreatySignatures.do, date accessed 9 August 2011.
83. 'The Antarctic Treaty' (1959), 1 December, http://www.ats.aq/documents/key-docs/vol_1/vol1_2_AT_Antarctic_Treaty_e.pdf, date accessed 9 August 2011.
84. T. Graham and D. J. LaVera (2003), *Cornerstone of Security: Arms Control Treaties in the Nuclear Era* (Washington, D.C.: University of Washington Press), p. 34.
85. Ibid., p. 35.
86. Jasentuliyana (1999), *International Space Law and the United Nations,* p. 33.
87. H. Blix et al. (2006), 'Weapons of Terror: Freeing the World of Nuclear, Biological and Chemical Arms', *Weapons of Mass Destruction Commission Report,* Stockholm, p. 147.
88. J. Dean (2005), 'The Current Legal Regime Governing the Use of Outer Space', in UNIDIR, 'Safeguarding Space for All: Security and Peaceful Uses', Conference Report, 25–26 March, http://www.unidir.org/pdf/articles/pdf-art2376.pdf, date accessed 9 August 2011, pp. 36–39.
89. D. Wolter (2006), *Common Security in Outer Space and International Law* (Geneva: United Nations UNIDIR), http://www.unidir.org/pdf/ouvrages/pdf-1-92-9045-177-7-en.pdf, date accessed 8 August 2011, p. 21.
90. Tronchetti (2011), 'Preventing the Weaponization of Outer Space', p. 83.
91. Cf. V. Kopal (2005), 'International Legal Regime on Outer Space: Outer Space Treaty, Rescue Agreement and the Moon Agreement', Presentation at the UN Office of Outer Space Affairs Workshop on Space Law, Abuja, Nigeria, 21–24 November, http://www.oosa.unvienna.org/pdf/sap/2005/nigeria/presentations/01-01.pdf, date accessed 9 August 2011.
92. H. Tuerk (2009), 'The Negotiation of the Moon Agreement', Space Law Symposium, Vienna, Austria, 23 March, http://www.unoosa.org/pdf/pres/lsc2009/symp00.pdf, date accessed 9 August 2011, p. 3.
93. Blix et al. (2006), 'Weapons of Terror: Freeing the World of Nuclear, Biological and Chemical Arms', p. 149.
94. 'Rescue Agreement' (1968), in United Nations (2002), *United Nations Treaties and Principles on Outer Space,* http://www.oosa.unvienna.org/pdf/publications/STSPACE11E.pdf, date accessed 9 August 2011, pp. 9–12.
95. 'Outer Space Treaty' (1967), Article 5.
96. 'History: Apollo 1', NASA website, http://history.nasa.gov/Apollo204/, date accessed 9 August 2011.
97. 'Deadly Accidents in the History of Space Exploration' (2003), *USA Today,* 2 January, http://www.usatoday.com/news/nation/2003-02-01-accident-timeline_x.htm, date accessed 9 August 2011.
98. Cf. V. Kopal (2005), 'International Legal Regime on Outer Space: Outer Space Treaty, Rescue Agreement and the Moon Agreement'.
99. Jasentuliyana, *International Space Law and the United Nations,* p. 34.
100. 'Liability Convention' (1972), in United Nations (2002), *United Nations Treaties and Principles on Outer Space,* http://www.oosa.unvienna.org/pdf/publications/STSPACE11E.pdf, date accessed 9 August 2011, pp. 13–21.

101. 'Convention on International Liability for Damage Caused by Space Objects', UNOOSA website, http://www.oosa.unvienna.org/oosa/en/SpaceLaw/liability.html, date accessed 9 August 2011.
102. Y. Zhao (2004), 'The 1972 Liability Convention: Time for Revision?' *Space Policy*, vol. 20, p. 117.
103. 'Convention on International Liability for Damage Caused by Space Objects', UNOOSA website.
104. 'Treaty signatures', UNOOSA website, http://www.oosa.unvienna.org/oosatdb/showTreatySignatures.do, date accessed 9 August 2011.
105. 'Liability Convention' (1972), Article 2.
106. Ibid., Article 1 (c).
107. Zhao (2004), 'The 1972 Liability Convention: Time for Revision?', p. 120.
108. 'Liability Convention' (1972), Article 7.
109. Zhao (2004), 'The 1972 Liability Convention: Time for Revision?', p. 121.
110. 'Registration Convention' (1976), in United Nations (2002), *United Nations Treaties and Principles on Outer Space*, http://www.oosa.unvienna.org/pdf/publications/STSPACE11E.pdf, date accessed 10 August 2011, pp. 22–6.
111. 'Treaty Signatures', UNOOSA website, http://www.oosa.unvienna.org/oosatdb/showTreatySignatures.do, date accessed 9 August 2011.
112. 'Outer Space Treaty' (1967), Article 13.
113. 'Registration Convention' (1976), Article 3.
114. 'Convention on Registration of Objects Launched into Outer Space', UNOOSA website, http://www.oosa.unvienna.org/oosa/SORegister/regist.html, date accessed 9 August 2011.
115. 'Notifications from States and Organizations (Launch Year 1976 to the present)', UNOOSA website, http://www.oosa.unvienna.org/oosa/SORegister/docsstatidx.html, date accessed 9 August 2011.
116. 'Registration Convention' (1976), Article 10.
117. I. Diederiks-Verschoor (1999), *An Introduction to Space Law* (The Hague: Kluwer Law International), p. 49.
118. 'Moon Agreement' (1984), in United Nations (2002), *United Nations Treaties and Principles on Outer Space*, http://www.oosa.unvienna.org/pdf/publications/STSPACE11E.pdf, date accessed 10 August 2011, pp. 27–35.
119. 'Moon Agreement' (1984), Article 3.
120. Cf. Kopal (2005), 'International Legal Regime on Outer Space: Outer Space Treaty, Rescue Agreement and the Moon Agreement'.
121. 'Agreement Governing the Activities of States on the Moon and Other Celestial Bodies', UNOOSA website, http://www.oosa.unvienna.org/oosa/en/SpaceLaw/moon.html, date accessed 9 August 2011.
122. Cf. F. G. Klotz (1998), *Space, Commerce and National Security* (New York: Council on Foreign Relations).
123. Diederiks-Verschoor, *An Introduction to Space Law*, p. 49.
124. 'Moon Agreement' (1984), Article 11 (5).
125. Ibid., Article 11 (7).
126. United Nations General Assembly (2009), 'Prevention of an Arms Race in Outer Space', A/C.1/64/L.25, 16 October, http://www.reachingcriticalwill.org/political/1com/1com09/res/L25.pdf, date accessed 9 August 2011.

127. United Nations General Assembly (2010), 'Transparency and Confidence-Building Measures in Outer Space Activities', A/C.1/65/L.38, 15 October, http://www.reachingcriticalwill.org/political/1com/1com10/res/L38.pdf, date accessed 9 August 2011.
128. A. D. Terekhov (1997), 'UN General Assembly Resolutions and Outer Space Law', *Proceedings of the 40th Colloquium on the Law of Outer Space/International Institute of Space Law of the International Astronautical Federation*, issue 40, p. 103.
129. Ibid.
130. United Nations General Assembly (1982), 'The Principles Governing the Use by States of Artificial Earth Satellites for International Direct Television Broadcasting', Res. 37/92, 10 December, http://www.oosa.unvienna.org/oosa/en/SpaceLaw/gares/html/gares_37_0092.html, date accessed 9 August 2011.
131. United Nations General Assembly (1986), 'The Principles Relating to Remote Sensing of the Earth from Outer Space', Res. 41/65, 3 December, http://www.unoosa.org/oosa/en/SpaceLaw/gares/html/gares_37_0092.html, date accessed 20 August 2011.
132. United Nations General Assembly (1992), 'The Principles Relevant to the Use of Nuclear Power Sources in Outer Space', Res. 47/68, 14 December, http://www.unoosa.org/oosa/en/SpaceLaw/gares/html/gares_47_0068.html, date accessed 20 August 2011.
133. 'Nation: Cosmos 954: an Ugly Death' (1978), *Time Magazine*, 6 February, http://www.time.com/time/magazine/article/0,9171,945940-1,00.html, date accessed 10 August 2011.
134. United Nations General Assembly (1992), 'The Principles Relevant to the Use of Nuclear Power Sources in Outer Space'.
135. United Nations General Assembly (1996), 'The Declaration on International Cooperation in the Exploration and Use of Outer Space for the Benefit and in the Interest of All States, Taking into Particular Account the Needs of Developing Countries', Res. 51/122, 13 December, http://www.unoosa.org/oosa/en/SpaceLaw/gares/html/gares_51_0122.html, date accessed 10 August 2011.
136. 'Strategic Arms Limitation Talks (SALT I)', Arms Control Association website, http://www.armscontrol.org/documents/salt, date accessed 9 August 2011.
137. The White House, Office of the Press Secretary (2010), 'Official Administration Explanation of New START agreement', Center for Arms Control and Non-Proliferation website, 25 March, http://armscontrolcenter.org/policy/nuclearweapons/articles/official_administration_explanation_of_new_start_agreement/, date accessed 9 August 2011.
138. 'Intermediate-Range Nuclear Forces Treaty', Arms Control Association website, http://www.armscontrol.org/documents/inf, date accessed 9 August 2011.
139. 'The Missile Technology Control Regime', MTCR website, http://www.mtcr.info/english/index.html, date accessed 9 August 2011.
140. 'MTCR Partners', MTCR website, http://www.mtcr.info/english/partners.html, date accessed 9 August 2011.
141. D. A. Ozga (1994), 'A Chronology of the Missile Technology Control Regime', *Nonproliferation Review*, vol. 1, issue 2, p. 66.

142. 'Guidelines for Sensitive Missile-Relevant Transfers', MTCR website, http://www.mtcr.info/english/guidetext.htm, date accessed 9 August 2011.
143. MTCR (2008), 'Missile Technology Control Regime. Equipment, Software and Technology Annex', MTCR/TEM/2008/Annex/001, 5 November, http://www.mtcr.info/english/MTCRTechnicalAnnexNov2008.pdf, date accessed 9 August 2011.
144. Ozga (1994), 'A Chronology of the Missile Technology Control Regime', p. 69.
145. Cf. A. Pikayev, L. Spector, E. Kirichenko and R. Gibson (1998), *Russia, the US and the Missile Technology Control Regime*, International Institute for Strategic Studies, Adelphi Paper 317 (Oxford: Oxford University Press).
146. Ozga (1994), 'A Chronology of the Missile Technology Control Regime', p. 71.
147. 'US Open to Dialogue with China on MTCR Related Sanctions' (1993), TV interview of Under Secretary of State Lynn Davis regarding the imposition of sanctions against China for its violation of the Missile Technology Control Regime, Federation of American Scientists' website, 25 August, http://www.fas.org/nuke/control/mtcr/news/930826-300844.htm, date accessed 9 August 2011.
148. Ozga (1994), 'A Chronology of the Missile Technology Control Regime', p. 69.
149. As of 30 June 2011, 133 countries have subscribed to the Hague Code of Conduct against Ballistic Missile Proliferation (HCoC), HCoC website, http://www.hcoc.at, date accessed 9 August 2011.
150. P. Kerr (2003), 'Code of Conduct Aims to Stop Ballistic Missile Proliferation', *Arms Control Today*, January/February, http://www.armscontrol.org/act/2003_01-02/icoc_janfeb03, date accessed 9 August 2011.
151. 'International Code of Conduct against Ballistic Missile Proliferation' (2002), 25 November, http://www.acronym.org.uk/docs/0211/doc13.htm#01, date accessed 9 August 2011, Article 3 (b).
152. Ibid., Article 4.
153. Kerr (2003), 'Code of Conduct Aims to Stop Ballistic Missile Proliferation'.
154. Hitchens (2010), 'Multilateralism in Space', p. 21.
155. D. Wright (2007), 'Space Debris from Antisatellite Weapons', *Bulletin of the Atomic Scientists*, 1 October, http://www.thebulletin.org/web-edition/features/space-debris-antisatellite-weapons, date accessed 9 August 2011.
156. Cf. F. G. Klotz (1998), *Space, Commerce and National Security*.
157. UNIDIR (2004), 'Safeguarding Space for All: Security and Peaceful Uses', Conference Report, 25–26 March, http://www.unidir.org/pdf/activites/pdf2-act254.pdf, date accessed 10 August 2011, p. 1.
158. United Nations (2007), 'Conference on Disarmament Hears Statements from 17 States on Prevention of an Arms Race in Outer Space', *Press Release*, 13 February, http://www.reachingcriticalwill.org/political/cd/press07/13Feb2007.html, date accessed 10 August 2011.
159. B. MacDonald, Presentation made at the Space Security 2009 Conference, United Nations Institute for Disarmament Research, Geneva, Switzerland 15–16 June, 2009, http://www.unidir.ch/pdf/ouvrages/pdf-1-92-9045-009-E-en.pdf, date accessed 10 August 2011.
160. Hitchens (2010), 'Multilateralism in Space', p. 22.

161. Ibid., p. 22.
162. Tronchetti (2011), 'Preventing the Weaponization of Outer Space', p. 86.
163. Ibid., p. 87.
164. N. Peter (2009), 'Space Power and Europe in the 21st Century', *European Space Policy Institute Perspectives*, issue 21, April, http://www.espi.or.at/images/stories/dokumente/Perspectives/ESPI_Perspectives_21.pdf, date accessed 27 July 2011, p. 2.
165. L. Grego and D. Wright (2010), *Securing the Skies: Ten Steps the United States Should Take to Improve the Security and Sustainability of Space* (Washington, D.C.: Union of Concerned Scientists), http://www.ucsusa.org/assets/documents/nwgs/securing-the-skies-full-report-1.pdf, date accessed 31 March 2011, p. 5.
166. Ibid., p. 1.
167. N. R. F. Al-Rodhan (1998), *The Three Pillars of Sustainable National Security in a Transnational World* (Zurich: LIT), p. 129.
168. Grego and Wright (2010), *Securing the Skies*, p. 1.
169. N. R. F. Al-Rodhan (2007), *The Five Dimensions of Global Security: Proposal for a Multi-sum Security Principle* (Berlin: LIT), p. 137.
170. M. Krepon, T. Hitchens and M. Katz-Hyman (2011), 'Preserving Freedom of Action in Space: Realizing the Potential and Limits of US Spacepower', in C. D. Lutes, P. L. Hays (eds), with V. A. Manzo, L. M. Yambric, M. E. Bunn, *Toward a Theory of Spacepower: Selected Essays* (Washington, D.C.: National Defense University Press), http://www.ndu.edu/press/spacepower.html, date accessed 6 June 2011, p. 399.
171. Ibid., p. 400.
172. Al-Rodhan, *The Three Pillars of Sustainable National Security in a Transnational World*, p. 12.
173. N. R. F. Al-Rodhan (2007), *Symbiotic Realism: A Theory of International Relations in an Instant and Interdependent World* (Berlin: LIT Verlag), pp. 66–135.
174. N. R. F. Al-Rodhan (2009), *Sustainable History and the Dignity of Man: A Philosophy of History and Civilizational Triumph* (Zurich: LIT), p. 294.
175. Ibid., pp. 271–95.
176. N. R. F. Al-Rodhan (2009), *Neo-Statecraft and Meta-Geopolitics: Reconciliation of Power, Interests and Justice in the 21st Century* (Berlin: LIT), p. 202.
177. N. R. F. Al-Rodhan (2008), 'Emotional Amoral Egoism': A Neurophilosophical Theory of Human Nature and its Universal Security Implications (Zurich: LIT), p. 65.

Index

Note: The page numbers in bold are the ones for which a (sub-)section or a chapter has been dedicated.

CPSIA information can be obtained at www.ICGtesting.com
Printed in the USA
LVOW080202211112

308275LV00006B/23/P

9 780230 314450